Ancestral Haplotypes

Our Genomes Have Been Shaped In The Deep Past

Edward J Steele

NU Publishing

Published by Nearurban LLC, 1603 LBJ Freeway,
Suite 280, Dallas, TX 75234
Tel (972) 974 7130, Fax (214) 393-4001
Email : admin@nearurban.net

First Edition – 2015 - 2nd Impression
ISBN 978-0-9864115-0-2

Steele, Edward, J, 1948 –
Ancestral Haplotypes
1. Evolution. 2 Hereditary. 3 Natural Selection.
1. Title
Manufactured and printed by CreateSpace.

Cover: Images adapted from Figure 4 Kulski JK et al 1999 J Mol Evol 49: 675-683 and http://en.wikipedia.org/wiki/Meteor_Crater

Contents

Note on Perth and Boston Groups : "The Perth group " includes those led by Emeritus Professor Roger L. Dawkins (UWA) during his past and present affiliations with The University of Western Australia (Royal Perth Hospital, Queen Elizabeth II Hospital), Murdoch University and the C.Y. O'Connor ERADE Village Foundation Inc. (CYO) incorporating the Immunogenetics Research Foundation (IRF). The work by "The Boston group" is that led by Professors Chester A. Alper and Edmond J. Yunis at Boston Children's Hospital, Harvard University and Department of Pathology, Harvard Medical School respectively.

About the Author

Edward J Steele was born and grew up in Darwin, NT, Australia. He gained his PhD from the University of Adelaide in 1976 under Professor Derrick Rowley for research on secretory IgA antibodies and intestinal immunity to cholera. Since then he has worked at a number of institutes and universities in Canada, UK and Australia. His main research theme has been the molecular mechanisms of the somatic and germline evolution of antibody variable genes for which he received the R.D. "Pansy" Wright Medal from the University of Melbourne (2009). He has also published or edited several books including *Somatic Selection and Adaptive Evolution* (1979), *Somatic Hypermutation in V-Regions* (1991) and *Lamarck's Signature* (1998). He is currently an Honorary Research Fellow with the C.Y. O'Connor ERADE Village Foundation, in Canning Vale, Perth, Western Australia.

Foreword

This book recounts a great scientific discovery. It tells the story of the leviathan effort to demonstrate that the structure of the DNA sequences in our chromosomes is not what it might seem. A simple view of evolution - from popular books and TV documentaries - has invaded main stream scientific thinking : that all our important genetic inheritance can be viewed under a simple Darwinian view of evolution based on recent and ongoing gene mutation followed by natural selection. To be more explicit, the neo-Darwinian paradigm based on population genetics and statistics.

The discovery of ancestral haplotypes is a heroic tale spanning almost 35 years documenting the efforts of the Australian group based in Perth - the IRF/CYO group - led by Professor Roger L Dawkins. It is a journey of scientific discovery which has uncompromisingly shown that in most if not all regions of our genome many of our important genes are very old, dating to the early migration of *Homo sapiens* out of Africa ca. 100,000 years ago. Many probably arose earlier in our primate ancestors. Large chromosomal blocks carrying tens to hundreds of "genes frozen in time" are shuffled at each generation to produce individual human beings who for the most part are healthy and robust and well adapted. These ancestral sequences circulate in the human population "gene pool" and occur at respectable frequencies (>1%). However the down side is that many people also carry genetic disease predispositions as a legacy of our primate and early human origins and thus appear not so lucky. However they usually survive in the most part to adulthood and have children of their own - their inheritance is part of a complex beneficial integrated adaptation which has been passed down over thousands of years from our ancestors. Nevertheless if we are to do anything positive about these apparent evolutionary aberrations we need to first understand the *most likely genetic story* of how we arrived at our present condition. This book tells this story as accurately as is possible - given the "strangeness" and complexity of the new genetic data concerning "ancestral haplotypes."

Acknowledgements

The author acknowledges the assistance of Sue Lester in the tracking down of papers and documents, and for transcribing the interview of Professor Roger L Dawkins, by Stuart Reid for the Oral History Project, National Library of Australia, 12th December 1993, Melaleuka Stud, Canning Vale, WA. I thank Drs Guy Grimsley and Joseph Williamson for reading and commenting on the draft, as well as Dr Sally Lloyd for assistance in various aspects of the writing. I particulary thank Dr Jerzy Kulski for his many insightful comments and suggestions on the draft manuscript. I also thank Mr Anthony Lloyd for critical discussions. I appreciate the time provided by Professor Roger Dawkins for interviews to clarify points of fact. Apart from this assistance the narrative of the history of the main scientific discoveries of the IRF/CYO group is that of the author who takes full responsibility for the emphasis and interpretation of the scientific and historical data. The AL and M Dawkins Foundation is acknowledged for support to attend CYO Foundation Quarterly Retreats during the writing of this book.

Preface

This book aims to be an informed and critical synthesis of the scientific work which has helped establish our current views of the DNA sequence structure of the human genome. It will focus on the research achievements of the Perth group led by Professor Roger L. Dawkins of the Immunogenetics Research Foundation (IRF) and the C.Y. O'Connor ERADE Village Foundation (CYO) over 35 years. Their work on the genomics of the major histocompatibility complex (MHC) has played a significant role in establishing the "Ancestral Haplotype" concept as an accurate description of the polymorphic frozen block structure of the human and wider mammalian genome. In different places it will also contain a corresponding account of the complementary clinical immunology and genetic studies on the MHC by the Boston Group led by Professors Chester A. Alper and Edmond J. Yunis on the same genetic regions within the MHC which they identify as Conserved Extended Haplotypes (CEHs). The author also acknowledges the wider contributions over many years of the international MHC multicentre laboratories who actively nurtured the work of the Perth and Boston groups through the International Histocompatibility Workshops. These established standard criteria and repositories for HLA typing antisera and HLA defined homozygous and heterozygous cell lines.

Rather than litter the text with the usual stigmata of a scientific paper key references are, in the main, added to "Further Readings" at the end of each Chapter. With respect to Figures and Tables I have also tried to keep them straightforward and use wherever possible freely available internet-based material to illustrate a point. In some cases data tables and figures are less clear in print as they have been adapted from historical materials.

EJ Steele, Canning Vale October 2014

Definition: Haplotype / Haplotypes

A haplotype is a group of genes within an organism that was inherited together from a single parent. The word "haplotype" is derived from the word "haploid," which describes cells with only one set of chromosomes, and from the word "genotype," which refers to the genetic makeup of an organism. A haplotype can describe a pair of genes inherited together from one parent on one chromosome, or it can describe all of the genes on a chromosome that were inherited together from a single parent. This group of genes was inherited together because of genetic linkage, or the phenomenon by which genes that are close to each other on the same chromosome are often inherited together. In addition, the term "haplotype" can also refer to the inheritance of a cluster of single nucleotide polymorphisms (SNPs), which are variations at single positions in the DNA sequence among individuals.

By examining haplotypes, scientists can identify patterns of genetic variation that are associated with health and disease states. For instance, if a haplotype is associated with a certain disease, then scientists can examine stretches of DNA near the SNP cluster to try to identify the gene or genes responsible for causing the disease.

Further Exploration
Concept Links for further exploration
pharmacogenomics | HapMap Project | genome wide association study | chromosome | DNA | gene | allele | genotype | Linkage | SNP | haploid

http://www.nature.com/scitable/definition/haplotype-haplotypes-142

****

Chapter 1

Prologue

Summary of Ancestral Haplotype Concept. State of Biomedical Scientific Knowledge until about 1980. General Background in the Wider Field of Genetics: Neo-Darwinism. Genetic Surprises : Ig V-D-J Recombination; Retroviruses and Retrogenes and Exon-Intron Gene Structure. Further reading.

As a biomedical scientist I am the writing this book for a biomedical readership interested in the major genetic issues of our day. That is, those medical disorders which affect many human beings and how we evolved to our present state. I have also written this historical story, in parts, in an argumentative style. This is not so much to offend but to drive home what I believe are the often ignored stand out features of genomic analysis with respect to disease predispositions and transplantation and how these data relate to us all.

Summary Of The Ancestral Haplotype Concept

All current evidence supports the view that the human and wider mammalian genome has a quantal ancestral

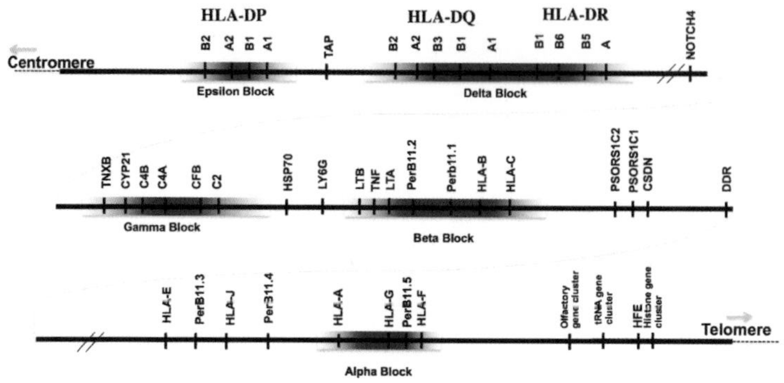

Figure 1 Ancestral Haplotypes and Polymorphic Frozen Blocks within The Human Major Histocompatibility Complex. Each ancestral haplotype has its own unique map of DNA sequence features which distinguish it from other ancestral haplotypes viz. single nucleotide polymorphisms (SNPs), copy number variations (CNVs), segmental duplications, insertion and deletion events (indels) including retroviral and retro-element like-insertions (RLEs). The full length DNA sequence shown in the figure is approximately 4 Mb. Recombination occurs far more frequently between the blocks than within and mutations within blocks are effectively suppressed. Adapted from Dawkins et al 1999 and the MHC Map (from Trowsdale and Knight 2013) as well as Lloyd S, Bayard D, Lester S, Williamson JF, and Dawkins RL (2013) The Value of Haplotyping. InterBull meeting, Nantes, France, August 2013.

structure, that is, highly complex and multifactorial polymorphic genetic information is packaged in large "quanta" on our chromosomes (~500Kb to ≥ 4Mb). The prototype data for this picture derives from many studies on the immune system's major histocompatibility complex (MHC) which clearly consists of polymorphic frozen blocks (PFBs) which are regions encoding long DNA ancestral haplotypes in the megabase range (Fig 1).

On the genomic scale, within such a long DNA sequence block, there maybe a small (≥ 10) or a very large number (≥ 200) of protein-coding genes and their associated regulatory regions (linked in tandem on the chromosome). Amongst these polymorphic sequences can be found segmentally duplicated regions, shorter insertion deletion events (indels), as well as more extended gene copy retroelements (REs) - which transpose via RNA template intermediates and reverse transcription events (RNA-to-DNA), such as SINEs and LINEs - and longer yet similar LTR-bounded endogenous retroviral elements (ERVs) as well as other REs (long retrotransposition elements) and DNA transposon events. Mutations within blocks are effectively suppressed and recombination occurs far more frequently between the blocks than witthin. Thus for a given block the polymorphism or variation inherent in such a long inherited sequence is akin to a classical "allelic" variant of a single copy gene, characterised, for example in the alleles defining differences from "wild type" such as the *wingless* or *white eye* mutations in the fruit fly, *Drosophila.*

Human genetic diversity and individuality is achieved by shuffling long polymorphic sequences. A linked string of genes and regulatory sequences formally mark an "ancestral haplotype" (AH). Each ancestral haplotype has its own characteristic DNA sequence.

When recombination takes place during meiosis in a diploid organism between paternal and maternal chromosomes (to produce haploid spermatozoa and ova) it occurs at the borders of such blocks. This model of the genome therefore predicts that recombination hot spots are located at the PFB borders as part of the conservation mechanism. Indeed all extant evidence is consistent with this prediction. Further, within a block, DNA repair mechanisms involved in the meiotic division of germ line cells are very efficient such that mutations - both simple point mutations and more complex ones such as insertion-deletion events (indels) - are rapidly repaired ("suppressed") as they emerge. The highly efficient mechanism of sister chromatid templated mismatch repair (MMR) is the most likely molecular mechanism ensuring this apparent "suppression." The variant sequences, or polymorphisms, defined by such blocks are thus considered as the true units of inheritance not individual genes as defined during the classical period (1900-1960) in genetic and population statistical studies - in fruit flies

(*Drosophila*), fungi (*Neurospora crassa*) and the more simple haploid bacteria and their viruses.

A side-by-side comparison of key conceptual differences between classical genetic ideas and the more complex genetic structures now uncovered in the higher mammalian and primate genome is shown in Table 1. In Box 1 are the conventional prescriptions proposed by population genetics under a pure neo-Darwinian

Box 1

Prescriptions Based on Population Genetics and Neo-Darwinism

• Genetic diversity or inherited variation requires on-going mutation.

• Diversity accumulates through errors in copying DNA.

• The unit of inheritance is the allele.

• At each locus, alleles may be deleterious, beneficial or neutral.

• Meiotic recombination scrambles maternal and paternal alleles.

• Linkage disequilibrium between SNPS define haplotypes.

From: Lloyd S, Bayard D, Lester S, Williamson JF, and Dawkins RL (2013) The Value of Haplotyping. InterBull meeting, Nantes, France, August 2013. An updated and complete exposition can be found in Hill (2014)

paradigm of evolution. This assumes ongoing mutation at discrete loci and free recombination, which, according to William Hill in a recent review, is " *the so-called "infinitesimal model" (of RA Fisher) ..formalized by Bulmer*

(1971, 1980), so properties of the normal distribution could be used: "normal theory presupposes an effectively infinite number of unlinked loci with infinitesimal effects; in consequence a finite change in the mean can be brought about by an infinitesimal change in gene frequencies" (Bulmer 1980, p. 150)."

Table 1 - Classical *versus* Complex Mammalian Genetic Systems

	Classical Genetics (Prokaryote Lower Eukaryotes)	**Complex Mammalian Genetics** (Polymorphic Frozen Blocks)
Unit Inheritance (Avg Size range)	Gene (Allele) or Cistron (100 bp to 5 Kb)	Ancestral Haplotype (100Kb - ≥4 Mb)
Number genes/loci involved	1	10s to 100s (includes regulatory regions such as ncRNA genes)
Genetic Diversity (Alternative forms)	Several alleles	Hundreds of alleles
Genetic Diversity (Major source)	Recent ongoing mutation	Inherited from Ancestors Created at Speciation Shuffling/Recombination of existing PFBs
Organisms that typify concept	Bacteria, bacteriophage *Drosophila, N. Crassa* Unicellular protozoans	Primate-Human lineage Higher mammalian genome (canine, bovine, murine)
Natural Selection?	Immediate and ongoing	At speciation-like events or major positive selection episodes
Applicability of Population Genetic Concepts?	Yes	No
Success/Recombination Linkage mapping?	Yes	No, too crude and imprecise because of long regions of suppressed recombination and mutation

The new genomic reality dictated by the ancestral haplotype structure of the genome thus runs counter to the more popular and accepted view of evolution that natural selection preserves recently arising mutations (Box 1). In classical genetics, according to RA Fisher, recombination occurs freely and at random between the genes such that in any population study the frequency of meiotic recombination between genes is a *direct linear measure* of the physical distance between the genes on that chromosome (Fig 2). Indeed, in *Drosophila,* a centimorgan (cM) is a unit defined by recombination frequency and is roughly equivalent to a length of 1 Mb of DNA (one million base pairs).

This book will document the scientific history of the discovery of ancestral haplotypes. It will provide plausible explanations on how they were generated - at least at the beginning of the human expansion about 100,000 years ago - and how they have been preserved or "frozen" ever since.

The author fully understands that these are very strange and challenging concepts for biomedical students and practitioners alike brought up through a traditional university education.

They contradict the more popular view of genetics outlined in the left hand column of Table 1 and promulgated by popular science writers such as Richard Dawkins and often amplified and reinforced in contemporary TV documentaries (however balanced and well meaning these try to be). This popular view as expected spills over into mainstream science and is promulgated as the "consensus" position.

Nothing could be further from the truth in this author's mind - so this history of the discovery of ancestral haplotypes will also be both argumentative and pedagogical by necessity.

The first key argument I put forward is this - since the time of RA Fisher modern genetics, and in particular Population Genetics, has depended on the assumption that random mutation occurs producing variant genes (alleles) and these recombine and segregate randomly in the population (via Mendelian inheritance). Thus a dominant allele (A) at a given locus will have a defined frequency in the population (p) and its alternative co-dominant allele (A') or recessive allele (a) will thus have a frequency 1- p (= q) which in an idealised randomly mating population such that the allele frequencies will be in Hardy-Weinberg equilibrium. A formal definition of the mathematics are readily found on the internet (or Box

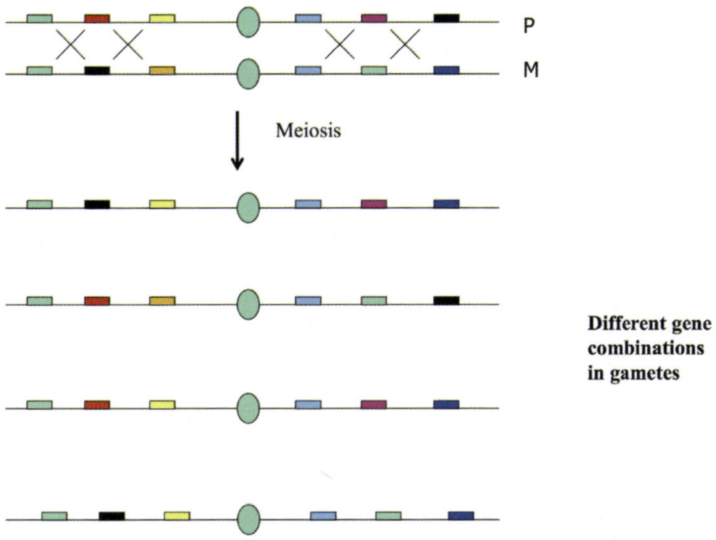

Figure 2 The "Genes as a Beads on a String" Model of Classical Genetics and Population Genetics (Neo-Darwinism). Schematic outline of classical genetics and Population Genetics (neo-Darwinism): genes envisaged as "beads on a string" with free and random recombination between loci during meiosis of paternal (P) and maternal (M) chromosomes. Some recombinant products in gametes are shown.

2 in Glossary). This mathematical way of describing genes and their population frequencies is applicable in context (Table 1) but nonsensical in the context of what is now known about the structure of the higher vertebrate genome.

Thus it is now clear that most if not all mammalian genes (and their regulatory *cis*-linked elements) *do not exist in isolation* and are not in free recombination as a reading of William Hill might suggest. First, most protein–coding genes (which occupy perhaps 5-10% of the DNA sequences in the genome) exist in tight clusters or "quanta". Second, alleles exist as gene variants *within* a frozen polymorphic ancestral block which is not subject to either recombination or further mutation - and this applies to single nucleotide polymorphisms (SNPs) as well as to more complex indels and gene copy number variations (GCNVs). This provides one of the reasons why MHC ancestral haplotypes are not detected by the SNP linkage dysequilibrium (LD) studies based on the HapMap consortium database (later in this book).

These facts about ancestral haplotypes undermine the fundamental assumptions of Population Genetics. The "quantal" genomic reality is at odds with conventional statistical analyses based on the bell curve or normal distribution (the Gaussian distribution) upon which population statistical genetics is based. If all SNPs, indels, micro- and mini-satellites etc. exist as part of a given ancestral block polymorphism then "allele" frequencies as such are misleading if not meaningless. Only "haplotype" frequencies should be cited and compared in this context.

Take as an example the established ancestral polymorphic frozen blocks located at the human major histocompatibility complex at chromosome band 6p21.3 (Fig 1). In Caucasoids, where most of the detailed work has been done, perhaps 70%-90% of all known diversity across the entire MHC region encompassing more than 200 protein coding genes and their regulatory regions *can be accounted for* by perhaps 50-200 polymorphic sequences and their recombinants thereof spanning the MHC region (~3.5 Mb to ≥ 4 Mb). So all genes, functional or non-functional, all existing SNPs or indels or CNVs at specified positions ("loci") in this region exist in the context of a defined, identifiable, *single contiguous ancestral DNA sequence.*

The reality of such contiguous sequences have to be *empirically determined* and confirmed by stable segregation in two to three generation family studies (at least). Statistics in the conventional sense does not enter into the equation - and if it does, it is to only define the frequency of an ancestral haplotype such as in a comparison with the frequency of other ancestral haplotypes. This departure from conventional Gaussian-based statistical thinking, and thus population genetical thinking, has far reaching consequences - it is not a trivial departure.

It will be shown in the course of this book that the detailed features of ancestral haplotypes worked out from the genomic structure of the MHC also applies at many polymorphic blocks (≥100) across the human genome. So our argumentative analysis is not restricted to the MHC and will be extended where necessary to illustrate a genomic pattern typical of the mammalian genome as a whole. The author notes there maybe one clear exception and special case - the V-to-D-to-J rearrangements at heavy and light chain immunoglobulin variable region (IgV) and T Cell receptor (TCR) loci - extreme germline specified polymorphism which sets up the somatic mechanism in the case of immunoglobulins for further antigen-driven hypermutation of rearranged V(D)J variable genes (and below).

State Of Biomedical Knowledge Up To About 1980

This book will also constitute an historical narrative. It is important here therefore to provide a short outline of the state of relevant scientific knowledge in the late 1970s pertinent to the discovery of polymorphic frozen blocks (Box 3). To anticipate a later tacit question, it is indeed surprising, given the concept of continuing "gene fluidity" discovered and made fashionable in that era – largely, as we show at the end of this book, a consequence of the "somatic cancer paradigm" - that a snap shot, the end result of such fluidity, could be frozen in time within an "ancestral haplotype" at speciation.

Modern immunology, and in turn molecular biology and genetics, were in their post-World War II infancy in the 1960s and 1970s. The double helix structure of DNA was discovered in 1953 by James Watson and Francis Crick. The DNA of a chromosome has two complementary polynucleotide strands (anti-parallel) and

Box 3 Genetic Rules Breakdown 1950s to 1980s

The accepted rules for genetics fashioned in simpler systems (bacteria, fruit flies) began to break down as geneticists began to analyse maize (Barbara McClintock) and genes in higher organisms (eukaryotes) and mammalian systems (molecular immunology - extreme MHC allelic polymorphisms and generation of antibody variable region diversity).

- Genes can be mobile not necessarily like fixed beads on a string
- Genes display extreme somatic diversity and germline polymorphisms
- Genes can rearrange position - programmed or aberrant (cancer)
- Genes are "split" and can be made of "bits"
- Genetic information can flow from DNA to RNA and back again
- RNA can be edited, processed and alternatively sliced

base paired such that pyrimidines (C,T) always pair by hydrogen bonding with purines (G,A): adenine bases pair with thymine bases via two hydrogen bonds (A:T) and guanine bases always pair with cytosine bases via three hydrogen bonds (G:C). Shortly after this was the discovery by Francis Crick and other colleagues of the triplet-based Genetic Code (Table 2) relating the nucleotide base sequence at the DNA and RNA level to the sequence of amino acids at the protein level – thus the articulation of molecular biology's famous dictum : The Central Dogma "that DNA gives RNA gives protein".

nonpolar polar basic acidic (stop codon)

Standard genetic code

1st base	2nd base U		2nd base C		2nd base A		2nd base G		3rd base
U	UUU	(Phe/F) Phenylalanine	UCU	(Ser/S) Serine	UAU	(Tyr/Y) Tyrosine	UGU	(Cys/C) Cysteine	U
	UUC		UCC		UAC		UGC		C
	UUA	(Leu/L) Leucine	UCA		UAA	Stop (*Ochre*)	UGA	Stop (*Opal*)	A
	UUG		UCG		UAG	Stop (*Amber*)	UGG	(Trp/W) Tryptophan	G
C	CUU		CCU	(Pro/P) Proline	CAU	(His/H) Histidine	CGU	(Arg/R) Arginine	U
	CUC		CCC		CAC		CGC		C
	CUA		CCA		CAA	(Gln/Q) Glutamine	CGA		A
	CUG		CCG		CAG		CGG		G
A	AUU	(Ile/I) Isoleucine	ACU	(Thr/T) Threonine	AAU	(Asn/N) Asparagine	AGU	(Ser/S) Serine	U
	AUC		ACC		AAC		AGC		C
	AUA		ACA		AAA	(Lys/K) Lysine	AGA	(Arg/R) Arginine	A
	AUG[A]	(Met/M) Methionine	ACG		AAG		AGG		G
G	GUU	(Val/V) Valine	GCU	(Ala/A) Alanine	GAU	(Asp/D) Aspartic acid	GGU	(Gly/G) Glycine	U
	GUC		GCC		GAC		GGC		C
	GUA		GCA		GAA	(Glu/E) Glutamic acid	GGA		A
	GUG		GCG		GAG		GGG		G

Table 2 The Genetic Code – Each triplet of RNA bases (where U is equivalent to T in DNA) is read from left (5') to right (3') and specifies the next amino acid in the sequence of triplets (= protein coding gene) to be incorporated into the growing protein chain. The synthesis of a polypeptide chain, or protein, occurs outside the nucleus in the cytoplasm as the messenger (m)RNA template is fed through the ribosomal machinery. The codon AUG both codes for the amino acid methionine and serves as an initiation site where translation into protein begins. See http://en.wikipedia.org/wiki/Genetic_code

By the mid to late 1970s genes could be cloned and expressed in target cells (bacterial and eukaryotic) and the DNA could be rapidly sequenced by the dideoxynucleotide chain termination and primer extension techniques developed by dual Nobel Prize winner Fred Sanger. Recombinant DNA technology using restriction enzymes (Werner Arber) and bacterial plasmids and bacteriophage cloning vectors became necessary technologies in genetic characterisation and discovery. The chemical and manipulative precisions of molecular biology were beginning to enter most molecular immunology laboratories. In Australia the father of modern immunology Frank MacFarlane Burnet's clonal selection theory of antibody formation had been articulated (1957) and the first formal proofs of the concept rapidly secured by his student Gus Nossal and then later (1969) by his colleague Gordon Ada.

As I was doing my PhD in the mid 1970s it was shown that antibody producing B cells could be immortalized as *in vitro* grown B cell tumour lines by the "hybridoma" technique developed by Cesar Milstein and his student George Kohler enabling highly specific "reagent grade" monoclonal antibodies to be developed for clinical and research purposes. Self *versus* non-Self discrimination in transplantation studies had been defined and human transplantation of hearts and kidneys had begun. T and B lymphocytes responsible for cell-mediated and humoral (antibody) immunity respectively had been described earlier by Jacques Miller and Graham Mitchell and colleagues at the Walter and Eliza Hall Institute in Melbourne, and by the mid 1970s the concept of MHC restricted recognition of virus-infected target cells by cytotoxic T cells was discovered and established by Peter Doherty, Rolf Zinkernagel and Rolf's PhD supervisor Bob Blanden at the John Curtin School of Medical Research at the ANU in Canberra (for MHC Class I peptide presenting structures to generate cytotoxic T lymphocytes). Later, helper/regulator T lymphocytes were shown to be activated by Class II (H-2/HLA) antigen presentation structures by Baruj Benacceraf and his post doctoral student David Katz in Boston and other groups such as Jacques Miller and colleagues at the Walter and Eliza Hall Institute in Melbourne.

An interesting historical aside of this period was the work of Dawkins and Zilko at Royal Perth Hospital. They contributed to the definition of T and B lymphocytes by showing that a sub-set of human T cells responsive to T cell mitogens could bind and form rosettes with sheep red cells thus allowing physical separation by centrifugation on ficoll gradients. Nowadays the flow cytometry procedure developed by the Lee Herzenberg group in the 1980s at Stanford is the routine analytical and preparative tool deployed.

The primary rules of tissue transplantation governed by the major histocompatibility complex in the mouse (H-2 complex) on murine chromosome 17 had been worked out earlier by Peter Gorer, George Snell and also later by the acquired neonatal transplantation tolerance studies of Peter Medawar, Leslie Brent and their colleagues at St Mary's Hospital Medical School at Paddington in London.

The practical necessities of blood transfusions, as well as the potential success in tissue and organ transplantation throughout the 1960s and 1970s, drove the research focus quickly to the human leucocyte antigens (HLA) encoded within the MHC on chromosome 6. The key players in this exciting era of human immunology were Jean Dausset and other colleagues such as Rose Payne, Walter and Julia Bodmer, Jon J. van Rood and A. van Leeuwen, Paul Terasaki, Bernard Amos, Roy Walford, Fritz Bach, Flemming Kissmeyer-Nielsen, Arne

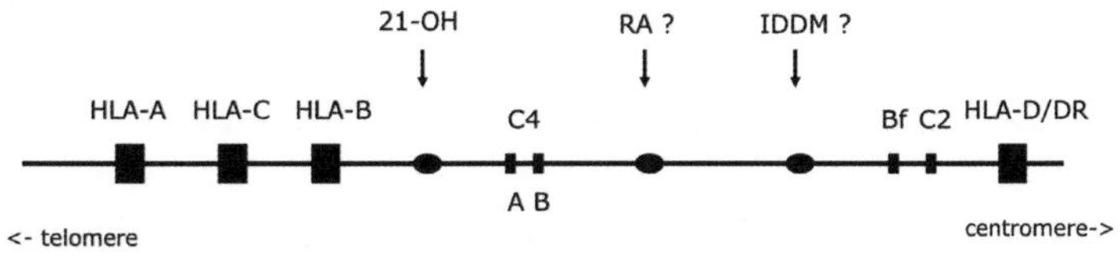

Figure 3 Early HLA Linkage Group Map of MHC ca. 1982. Adapted from O'Neill G.J., and Nerl C. (1982) The HLA-linked Complement Markers in Disease. In "Immunogenetics in Rheumatology : Musculoskeletal Disease and D-Penicillamine" Edited by R. L. Dawkins, F.T. Christiansen and P.J. Zilko. Pp24-28. Excerpta Medica. Amsterdam-Oxford-Princeton

Svejgard, J. Richard Batchelor and others. Many highly specific HLA tissue typing antisera defining the Class I and Class II antigens and other HLA serologically defined cell surface antigens or soluble gene products (complement components C4, C2, Bf) were exchanged and standardized. These specific serological reagents together with family studies for disease associations led to the early linkage map of the HLA markers making up the MHC (Fig 3).

Background In The Wider Field Of Genetics : Neo-Darwinism

The simple view of Population Genetics and genetic interpretation in general (which still holds sway) is outlined in Table 1, Box 1 and Fig 2. The author, not being a population geneticist, understands that statistics has its place in the biomedical sciences but not to the exclusion of biological understanding which includes the discovery of molecular mechanisms through empirical experimentation and biological analysis. As a graduate and postgraduate student in the 1960s and 1970s the author remembers full well the impact of Population Genetics on the teaching of genetic concepts in biology and medicine at that time.

It is now clear to this author that the acceptance of the discovery of ancestral haplotypes as a more accurate description of the polymorphic block structure of the human and mammalian genome has been *actively retarded* by the contemporary Population Genetics paradigm. This view still has a profound influence not only in the teaching of genetical concepts but in the way research is conducted. Quite frankly its power needs to be curbed if mankind is to make progress in genomic genetic analyses and in the genetic improvement of our livestock herds.

The most significant failure of Population Genetics in the post HUGO genome era (viz. the drafting of the first DNA sequence of the human genome in the year 2000) has been its inability to detect - as part of the HapMap project consortium - through SNP-Linkage Disequilibrium analyses (SNP/LD), the numerous long (≥ 3-4 Mb) polymorphic ancestral haplotypes discovered in the 1980s and 1990s by the Perth and Boston groups. This failure underpins the critical distinctions contrasted in Table 1. It will be dealt with again in a later chapter.

Genetic Surprises

There is no doubt the 1970s bore witness to three sets of dramatic discoveries in molecular and cell biology which completely undermined the tenets of Population Genetics and conventional evolutionary theory (although it was not recognised by the neo-Darwinian population geneticists at the time or since).

Immunoglobulin V-to-D-to-J Recombination

Within molecular immunology the somatic *versus* germline debate on antibody (immunoglobulin, Ig) diversity was still raging led by Melvin Cohn (somatic) and Leroy Hood (germline). Then Susumu Tonegawa dramatically showed in 1974 that antibody variable (V) genes were in bits in the germline and assembled during somatic life into a complete variable region gene (so called VDJ or VJ rearrangements). By the end of the decade the resolution, for mice, was as follows :

- There are a limited number of germline unrearranged Ig V elements.
- These somatically rearrange to D/J joining and diversity elements in the B cell DNA (combinatorial association) such that for a heavy Ig chain if there are 100 V elements, 10 D elements, 5 J elements, there are 100 x 10 x 5 = 5000 possible productively VDJ rearranged H chains - and if you include nucleotide additions at the joints then this number could be increased by a factor of 5 or 10.
- For the rearranged light chain (L) there would be 100 x 5 = 500 possibilities which if all the LJ joints were productive this would give a combinatorial number for HL variable region dimers of

500 x 5000 or 2.5 x 10^6 potential antibody specificities.

• In addition there is the somatic hypermutation mechanism in the Germinal Centre recruited to the VDJ rearrangement to increase the rate of nucleotide variation in antigen-stimulation of B cells. With somatic hypermutation superimposed on the combinatorial process there is a vast substrate of possibilities for appropriate affinity maturation during an antibody response to antigen.

A picture like this, with minor modifications, now applies to human Ig loci, as well as T cell receptor (TCR) loci (except that T cells driven by viral and other intracellular parasite antigens, do not normally somatically mutate apart from errors via nucleotide additions and deletions at the V-D-J and V-J joints).

Therefore, combinatorial association of smaller genetic regions for both the immunoglobulin (Ig) and the T cell receptor system (TCR) solved the problem of the adaptive immune response to unexpected foreign antigens. The immune system does not need to store in the species germline an impossible large number of encoded antigen specificities - it simply had to have an *inherited strategy* for the generation of diversity (GOD). That is, a limited number of germline V gene bits and their joining regions augmented in the case of antibodies by somatic hypermutation of the VDJ region.

Jumping ahead towards the end of this book we will see that on a genomic scale the Perth group through Roger Dawkins has recently discovered there must exist a genomic generator of diversity (GOD) likely to be centred in part on the multifactorial and polymorphic ancestral haplotypes of the MHC region. The MHC assists the reliable determination of the Self *versus* Non-Self discrimination in an individual (instance transplantation tolerance or rejection). Roger Dawkins and coworkers now hypothesize that it may well control the combinatorial outcome of the meiotic shuffling of polymorphic ancestral haplotypes across the genome at numerous blocks (>100). So we are now understanding better that there is likely to be an *inherited strategy* for human genetic individuality based on combinatorial association of conserved pre-tested polymorphism as well.

Retroviruses, Retrogenes, Retroelements.

These were totally unexpected genetic surprises to the scientific community (although foreshadowed by Howard Temin theoretically in 1959). Thus in 1970 Howard Temin and David Baltimore independently reported the discovery of reverse transcription in RNA tumour viruses. This was a dramatic modification to the accepted Central Dogma of Molecular Biology viz. DNA->RNA-> Protein. Francis Crick however was not surprised as he had always considered it as a logical option in nucleic acid information transfer. Temin and Baltimore showed that genetic information could flow back from RNA to DNA. Thus, base modified RNA molecules, could in theory introduce new genetic information from "the outside" back to DNA. Within a few years numerous viral and non-viral RNA sequences (retrogenes and retroelements) such as SINEs (Alu)/LINEs were lighting up as DNA copies all over the genome. It now turns out that many of these retroelements (REs) and retroviral (RV) integration events occurred thousands if not millions of years ago in the primate linage leading to humans and other mammalian groups (they are typical of genomes in the Plant Kingdom as well). Each species line has its own unique set of complex RE/RV integration signatures. Almost 45% of the DNA in higher mammalian genomes (primates-humans, dog, cow, mouse and rat) is composed of DNA copies of former RNA sequences (retroviruses, short interspersed nuclear elements (SINEs, such as Alu, tRNAs, B1 elements) and reverse transcriptase expressing long interspersed nuclear elements (LINEs) as well as functional and non-functional endogenous retroviral elements and their fragments (ERVs). We will discuss later in this book that at many places the genome is also extensively segmentally duplicated in much larger sections (> 10Kb if not > 100Kb).

Thus genes could jump and integrate elsewhere (e.g. retrotranspose) in the genome (Box 3 and Glossary). These discoveries also confirmed the work on mobile genetic elements by Barbara McClintock in maize in the 1940s and early 1950s. It is now clear at least in the human and mouse genome (separated by about 90 million years since the last common ancestor) that integration patterns of SINEs, LINEs and ERVs are highly non-random. In the human genome Alu retroelements integrate in gene rich regions and, in particular, more than 70% of such integration events favour introns (and below). Pathogenic retroviruses such as HIV-1 also integrate preferentially in actively expressed gene-rich regions and can become integrated and inherited vertically morphing over evolutionary time into more benign endogenous retroviruses (ERVs). In the considered opinion of Roger Dawkins and colleagues the insertion of such REs in genes and their surrounding flanks has had *major influences* leading to polymorphism in gene function, expression and regulation.

Temin and Baltimore's 1970 discovery of reverse transcription was prescient - it prepared the world for the HIV/AIDS pandemic which exploded without warning in 1981. It quickly allowed the biomedical community to figure out that the acquired immune deficiency disease was caused by a pathogenic retrovirus which probably jumped from a higher primate to man. The discovery of ERVs and retro-like elements (RLEs) is relevant to our later discussion on how polymorphic ancestral haplotypes are laid down during speciation and speciation-like events.

Exon-Intron Structure of Eukaryotic Genes

Up until the mid 1970s the DNA structure of genes was dominated by the paradigm of bacterial genetics and its

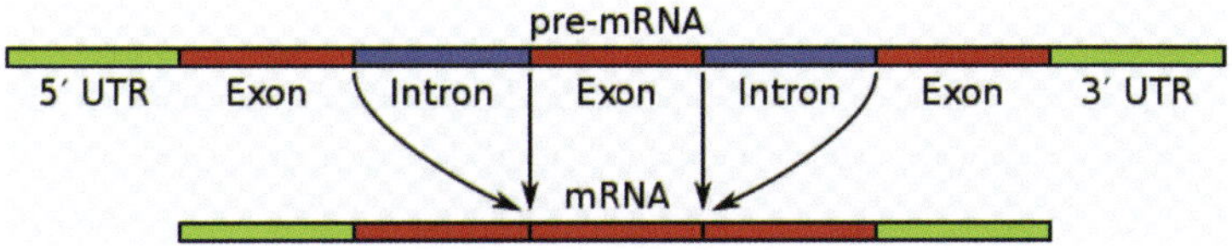

Figure 4 Exon-intron Structure of Eukaryotic Genes.
See http://en.wikipedia.org/wiki/File:Pre-mRNA_to_mRNA.svg

associated molecular biology (the Paris group of molecular biologists, Francois Jacob, Andre Lwoff, Jacque Monod and their colleagues). Thus a protein coding gene in bacteria (cistron) or a series of linked related genes involved in a biochemical pathway (polycistron, regulated as an "operon") were understood in a straightforward way. Embedded in the DNA specifying a particular gene were start sites for transcription (DNA copied to RNA) to produce the messenger (m)RNA and just down stream was the start site for protein synthesis (translation) located at the initiating codon "ATG" (AUG in mRNA, Table 2). Between the start and the end of the mRNA message the genetic information (triplet code) specifying a protein sequence was contiguous with the DNA. And the regulation of the expression of such genes was via "operator" sites and "repressor" proteins which bound the DNA at the 5' side of the gene where transcription of mRNA begins.

Then in 1977 all this changed when Phillip Sharp and Richard Roberts independently discovered "introns" in adenoviruses : the protein-coding genes of higher genetic systems were literally split - that is they had an exon (coding) and intron (non-coding) structure (Figure 4). Very soon almost all eukaryotic protein-coding genes were shown to have a similar intron-exon structure. Thus the copying of a gene by transcription into mRNA first produced a very long pre-mRNA transcript which then had to be processed and edited to remove the non-coding introns prior to moving to the ribosomes in the cytoplasm for translation of the spliced message into a amino acid sequence specifying a specific protein. The process of intron removal and other pre-mRNA processing events to produce a mature mRNA message turned out to be very complex. A key player is the "Spliceosome" complex of catalytic proteins and small RNAs which consists of maybe 200 of more molecular components which recognize and cut exon-intron junctions and then rejoin spliced exons to produce a contiguous mRNA.

Therefore as the 1970s came to a close it was clear the genes in higher organisms did not appear to behave in the simple way demanded by more rigid views of inheritance and genetic diversity.

Thus, in summary, the lymphocytes producing antigen recognition structures in the immune system could, in principle, respond to potentially produce millions of different antigen-specific receptors (B cell antibodies, T cell receptors). In principle, genes in higher mammalian and eukaryotic cells in general cells could be made into RNA and which then could potentially be reverse transcribed back to DNA prior to integrating back into the genome - thus augmenting by orders of magnitude the potential for genetic diversification both during somatic life (e.g. somatic hypermutation of Ig genes and unregulated in cancer) and during evolution (speciation and speciation-like events). Finally, genes in higher cells had a clear exon-intron structure vastly complicating the simple view of gene structures produced over the previous 10-15 years from bacterial genetics. Furthermore, the study of exon-intron structures soon moved on to the demonstration of alternative splicing strategies in numerous eukaryotic genes thus vastly increasing the potential information content of an organism's "transcriptome" (the universe of protein encoded spliced mRNA molecules).

In this context it is not surprizing that intronic (and intervening) sequences were regarded as “junk DNA” of no relevance to evolution. In fact we now know, and as we will see later, these sequences usually encode sites that when disrupted by retroviruses and retro-like elements can profoundly affect gene regulation. The ideas discussed in the pages of the journal *Nature* in the early 1980s that "junk DNA" plays no genetic role and is simply a manifestation of DNA behaving as a "selfish replicator" as put by Richard Dawkins would be contradicted and overturned by the 1990s.

It is clear then that a very comfortable linear view of gene structure and function was being overturned in the 1970s. It was into this environment that the first concrete hints began to emerge that "ancestral haplotypes" would violate some key rules of classical genetics. The clinical immunology laboratories in Perth and Boston began to show, first tentatively, and then more comprehensively, that the human genes encoding blood and tissue histocompatibility and general immune functions within the MHC were not following the expected rules of inheritance.

Further Reading Chapter 1

Ada, G.L. and Nossal, G. (1987) The Clonal Selection Theory. Scientific American. 255: 62.

Alper, C.A., Larsen, C.E., Dubey, D.P., Awdeh, Z.L., Fici, D.A. and Yunis, E.Y. (2006) The haplotype structure of the human major histocompatibility complex. Human Immunology 67: 73-84.

Altshuler, D., Brooks, L.D., Chakravarti, A., Collins, F.S., Daly, M.J. and Donnelly, P. (Members of writing group: The International HapMap Consortium.) (2005) A haplotype map of the human genome. Nature 437: 1299 - 320.

Baudat, F., Buard, J., Grcy, C., Fledel-Alon, A., Ober, C., Przeworski, M. et al. (2010) PRDM9 is a major determinant of meiotic recombination hotspots in humans and mice. Science 327: 836 - 840.

Billingham, R.E., Brent, L. and Medawar, P.B., (1956). Quantitative studies on tissue transplanataion immunity III Actively acquired tolerance. Philos. Trans.R.Soc. Lond. Series B 239: 357-414.

Blanden, R.V., Rothenfluh, H.S., Zylstra, P., Weiller, G.F. and Steele, E.J. (1998) The signature of somatic hypermutation appears to be written into the germline IgV segment repertoire. Immunological Reviews 162 : 117-132.

Brack, C. and Tonegawa, S. (1977) Variable and constant parts of th immunoglobulin light chain gene of a mouse myeloma are 1250 nontranslated bases apart. Proc. Natl Acad. Sci. (USA) 74: 5652 - 5656.

Brack, C., Hirama, M., Lenhard-Schuller, R. and Tonegawa, S. (1978). A complete immunoglobulin gene is created by somatic recombination. Cell 15 : 1-4.

Burnet, F.M. (1959) The Clonal Selection Theory of Acquired Immunity. Cambridge University Press, Cambridge, UK.

Cheung, V.G., Sherman, S.L. and Feingold E. (2010) Genetic control of hotspots. Science 327:791–792.

Cohn, M. (1974) The molecular biology of expectation. In Nucleic Acids in Immunology, Plescia, O.J. and Braun, W. Eds. Springer-Verlag, N.Y. p 671.

Crick, F.H. (1970) Central Dogma of Molecular Biology. Nature (Lond). 227 : 561-563.

Cunningham, A.J. Editor (1976) The generation of antibody diversity: A new look. Academic Press, London.

Davis, B.D., Dulbecco, R., Eisen, H.N., Ginsberg, H.S., Wood, W.B. and McCarthy , M. (1973) Microbiology, Harper and Row, NY.

Dawkins, R.L., Leelayuwat, C., Gaudieri, S., Tay, G., Hui, J., Cattley, S., Martinez, P. and Kulski J.K. (1999). Genomics of the major histocompatibility complex: haplotypes, duplication, retroviruses and disease. Immunological Reviews 167: 275-304.

Dawkins, R.L. and Zilko, P.J. (1975) Separation of cells involved in phytohaemagglutinin-induced mitogenesis and cytotoxicity. Nature 254 : 144-145.

Degli-Esposti, M.A., Leaver, A.L., Christiansen, F.T., Witt, C.S., Abraham, L.J. and Dawkins, R.L. (1992). Ancestral haplotypes: conserved population MHC haplotypes. Human Immunology 34: 242-252.

Doherty, P.C., Blanden, R.V. and Zinkernagel, R.M. (1976) Specificity of virus-immune effector T cells with H-2K or H-2D compatible interactions: Implications for H-antigen diversity. Trans. Rev. 29: 98-124.

Gabriel, S.B., Schaffner, S.F., Nguyen, H., Moore, J.M., Roy, J., Blumenstiel, B., et al. (2002) The structure of haplotype blocks in the human genome. Science 296: 2225-2229.

Goldfarb, T. and Lichten, M. (2010) Frequent and Efficient Use of the Sister Chromatid for DNA Double-Strand Break Repair during Budding Yeast Meiosis. PLoS Biol 8(10): e1000520. doi:10.1371/journal.pbio.1000520

Henderson, M (2008) *50 genetic ideas. You really need to know*. Quercus Publishing Plc, London.

Hill, W.G., Applications of population genetics to animal breeding, from Wright, Fisher and Lush to genomic prediction. Genetics, 2014. 196 (1): 1-16.

Hood, L. and Talmage, D. (1970). Mechanism of antibody diversity: Germline basis for variability. Science 168: 325-s334.

Hood, L.E., Weismann, I,L., Wood, W.R. and Wilson, J.H. (1984). Immunology 2nd Edit. Benjamin/Cummings Series in the Life Sciences. Addison-Wesley

Hozumi, N. and Tonegawa, S. (1976). Evidence for somatic rearrangement of immunoglobulin genes coding for variable and constant regions. Proc. Natl. Acad. Sci. (USA) 73: 3628-3632.

Klein, J. (1975) Biology of the Mouse Histocompatibility-2 Complex. Principles of Immunogenetics Applied to a Single System. Springer-Verlag, New York

Li, H., Cui, X., Pramanik, S. and Chimge, N-O. (2002) Genetic diversity of the human immunoglobulin heavy chain VH region. Immunological Reviews 190: 53-68.

Lindley, R.A. and Steele, E.J. (2013) Critical analysis of strand-biased somatic mutation signatures in TP53 versus Ig genes, in genome -wide data and the etiology of cancer ISRN Genomics. Vol 2013 Article ID 921418, 18 pages (Open Access online)

Martini, E., Borde, V., Legendre, M., Audic, S., Regnault, B., et al. (2011) Genome-Wide Analysis of Heteroduplex DNA in Mismatch Repair–Deficient Yeast Cells Reveals Novel Properties of Meiotic Recombination Pathways. PLoS Genet 7(9): e1002305. doi:10.1371/journal.pgen.1002305

McLure, C.A., Hinchliffe, P., Lester, S., Williamson, J.F., Millman, J.A., Keating., P.J., Stewart, B, J., and Dawkins, R.L. (2013) Genomic evolution and polymorphism: Segmental duplications and haplotypes at 108 regions on 21 chromosomes. Genomics 102 : 15-26

Myers, S., Bowden, R., Tumian, A., Bontrop, R.E., Freeman, C., MacFie, T.S., et al. (2010) Drive against hotspot motifs in primates implicates the PRDM9 gene in meiotic recombination. Science 327: 876-877.

Parvanov, E., Petkov, P.M. and Paigen, K. (2010) Prdm9 controls activation of mammalian recombination hotspots. Science 327: 835.

Ridley, M. 1993 Evolution. Blackwell Scientific Publications (3rd edition 2003).

Schroder, A.R.W., Shinn, P., Chen, H., Berry, C., Ecker, J.R. and Bushman, F. (2002) HIV-1 integration in the human genome favors active genes and local hotspots Cell 110: 521-529

Shapiro, J.A. (2011) Evolution: A View From The 21st Century. FT Press Science ISBN 978-0-13-278093-3

Steele, E.J. (Editor) (1991) Somatic Hypermutation in V-Regions. CRC Press. Inc. Boca Raton, FL

Steele, E.J. (2009) Lamarck and Immunity: Somatic and germline evolution of antibody genes. J. Roy. Soc. WA 92: 437-446.

Tonegawa, S. (1983) Somatic generation of antibody diversity. Nature 302: 575-581.

Stumpf, M.P.H. (2002) Haplotype diversity and the block structure of linkage disequilibrium. Trends Genet 18: 226 - 228.

Trowsdale, J. and Knight, J.C. (2013) Major Histocompatibility Complex Genomics and Human Disease Annu. Rev. Genomics Hum. Genet. 14: 301-323

Wall, J.D. and Pritchard, J.K. (2003) Haplotype blocks and linkage disequilibrium in the human genome. Nat Rev Genet 4: 587-597.

Watson, J.D (1965) Molecular Biology of the Gene. W.A. Benjamin. Inc. 1st edit.

****

Chapter 2

Inherited Complex Diseases :Supratypes or Haplotypes?

IRF/CYO Research in the late 1970s to 1982

Supratypes versus Haplotypes? HLA and Disease Associations. Family Studies: Supratypes, Complotypes, Haplotypes.

Supratypes *versus* Haplotypes?

Paul Terasaki a pioneer of HLA research reflected in 1990 on the early years, the 1950s and 1960s:

> " My interest in HLA stemmed from its possible application to transplantation. Though the implycations of HLA were far reaching, far beyond applications to transplantation,... progress in HLA was fuelled principally by its potential application to transplantation. Scientists do not live in a vacuum, they need funds to carry out their research, ...almost all grant requests for the study of HLA included some form of statement that knowledge of HLA will contribute to society through the improvement of transplantation."

While transplantation was and still is the *raison d'etre* of HLA research it became increasingly clear that the 'major histocompatibility locus' which encoded the serologically defined HLA surface antigens was also a central player in numerous other life threatening and debilitating diseases. However, as we will see later in this book, the discovery of the various polymorphic ancestral haplotypes encompassing the MHC locus has also led to potential great improvements in the success of transplantation, particularly such life-enhancing procedures as bone marrow transplantation. It has come to pass, yet not widely appreciated by the biomedical community, that if donors and recipients can be matched for the most important regions of their MHC haplotypes i.e. across the region from HLA-B to DR, then long term graft survival is greatly increased. This direct improvement to human health care and well being has arisen from the clinical immunology research of the Perth and Boston groups.

HLA and Disease Associations

As the 1970s came to a close there were reports appearing at an ever increasing rate of associations between immunogenetic markers encoded in the MHC and genetic predisposition to disease.

Accordingly, in April 1982 Roger Dawkins as Head of the Departments of Clinical Immunology at Royal Perth Hospital and The Queen Elizabeth II Medical Centre in Perth convened two Workshops (The "Rottnest Island" meeting). The aim was to focus discussions on some of the genetic and statistical principles underlying the observations. In this way it was hoped that the apparent discrepancies in the data and interpretations between the different laboratories could be ventilated and resolved. The collected discussions of the workshops were published in a volume entitled "Immunogenetics in Rheumatology : Musculoskeletal Disease and D-Penicillamine". The immunogenetics of the side effects to drugs such as D-Penicillamine were reported in a prior larger workshop. The Workshop participants were asked to prepare commentary of their data to contribute to a

"State of the Art" volume. Pretty well all invited participants in Australia and overseas complied with Roger Dawkins' invitation.

An array of disease associations with allelic markers within the MHC were reported including : Multiple Sclerosis (MS) ; Juvenile Diabetes Mellitus (or Insulin Dependent Diabetes Mellitus, IDDM) ; Psoriasis; Coeliac Disease (CD) ; 21 Hydroxylase Deficiency (21-OH, Congenital adrenal hyperplasia) ; Ankylosing Spondylitis (AS) ; Autoimmune Diseases such as Systemic Lupus Erythematosus (SLE) ; Myasthenia Graves (MG) and Rheumatoid Arthritis (RA). They appeared genetically linked to allelic variants of particular HLA cell surface antigens or soluble proteins of the Complement system such as C4, properdin factor B (Bf) and alleles of C2 (detected by immunoelectrophoresis). An early map of the HLA linkage group is shown in Figure 3 indicating that the Complement encoding markers belong as a cluster located between the telomeric HLA-B Class I locus and centromeric HLA-D/DR Class II loci.

The term “haplotype” as formally defined by the journal *Nature* is in the definition Box of the front pages. However it was the great Italian immunologist Ruggero Ceppellini and his coworkers who provided the initial definition of a haplotype when they recognised in the mid 1960s that sets of serological patterns of leukocyte antigens segregated in families. Then in the mid to late 1970s the importance of potential extended or ancestral HLA haplotypes was demonstrated by the Boston group of Alper and Yunis with the haplospecific copy number variations of the C4 complement gene within the central region of the MHC. Each putative ancestral haplotype identified appeared to have a specific C4 copy number which the Perth group then showed related to serum concentration of C4 and susceptibility to disease. For example, complement allotyping revealed the association between C4A null and SLE.

Family Studies: Supratypes, Complotypes, Haplotypes

However all efforts in this period to pin down a particular disease to a particular HLA polymorphic allele were, with some rare exceptions (HLA-B27 and Ankylosing Spondylitis, AS) unsuccessful.

It was the results of the nuclear family studies which both shed light on this lack of success and also deepened the problem of data interpretation. Thus it was becoming apparent that if a disease proband was HLA typed and complotyped along with all available siblings and parents (and grand parents if available) interesting yet fragmentary linked patterns of inheritance began to emerge. A given disease appeared to be associated with various linked patterns of markers associated with a number of different HLA and complement loci within the MHC. These early data and their interpretation appeared in a major invited review by the Perth group in 1983 entitled "Disease associations with complotypes, supratypes and haplotypes".

In the author's opinion this is an important paper as it laid out all the issues clearly as well as the uncertainties and set the conceptual research agenda for the next decade.

Thus a formal distinction was made between the genetic descriptive term "haplotype" and other linked alleles (associated with disease) they termed "supratypes". The Perth group went to considerable lengths to explain the differences. If a chromosomal segment was inherited *en bloc* from one parent they recommended that the term "haplotype" be used. The assumption had to be made there was no recombination. Thus in a family study such a combination of alleles at HLA-A, HLA -B and HLA-DR can then be derived and haplotypes making up the genotypes assigned. However the uncertainty was that the unknown alleles in intervening loci could still differ and unless extensive inheritance patterns could be established this may still not imply identity when the same combination cropped up in another family study.

Thus the need for another term to cover the apparent linkage of markers yet communicate the uncertainty about the intervening regions :

> "The term supratype (Dawkins et al. 1982) is used to refer to combinations of selected alleles at different loci without implying that the intervening alleles are necessarily identical. Supratypes will include haplotypes deduced from one or more family studies but may also be used to describe combinations of alleles recognised from a phenotype." (Dawkins et al 1983).

However in many situations family data was not available. For the first few years of the decade therefore the term "supratype" was used to describe these repeating patterns of apparently linked MHC alleles associated with various chronic diseases.

The term “supratype” became the provisional name for an extended MHC “haplotype” in unrelated individuals and in the absence of family segregation data. This precaution was considered justified given the history at that time of inappropriate usage of the term "haplotype". The other unstated issue was that a particular combination of Class I, II, and Complement alleles could have been the result of recombination of different alleles in the intervening sequences and brought together and maintained by Darwinian natural selection.

Although this begs the question of why an apparently maladapted linked set of genes should be maintained in human evolution. (For those interested readers I include at the end of this book more detail of this early history from an interview and discussion with Roger L Dawkins.)

Designation	A	Cw	B	Bf	C2	C4A	C4B	DR	Disease
8/3-0-1	1	7	8	S	C	Q0	1	3	MG, SLE, IDDM
14/1-2-2	–	–	14	–	–	2	2	1	Late 21-OH°
18/3-3-0	–	–	18	F1	C	3	Q0	3	IDDM
18/2-4-2	25	–	18	S	Q0	4	2	2	C2 deficiency
44/4-3-0	2	5	44	S	–	3	Q0	4	D-Pen thrombocytopenia
47/7-0-3	–	–	47	F	–	0	3	7	Early 21-OH°
62/4-3-2.9	2	3	62	S	C	3	2.9	4	IDDM, RA
62/4-4-2	–	–	62	S	B	4	2	–	?

Table 3 Some Caucasian supratypes of interest in HLA disease associations. Adapted from Table II Dawkins et al 1983 Q0=quantitatively zero.MG, Myasthenia Gravis; SLE, Systemic Lupus Erythematosus; IDDM, Insulin Dependent (Type I) Diabetes Mellitus; 21-OH°, 21 Hydroxylase deficiency ; C2 deficiency, C2 deficiency; sD-Pen thrombocytopenia, D-Penicillamine induced Thrombocytopenia; Early and late 21-OH°, Early and late onset 21-OH° ; RA, Rheumatoid Arthritis.

An example of the data produced in this period is shown Table 3. Note that in the left hand column the group designation of the "supratype" is given. It is usually identified first by the allele at the telomeric HLA-B locus, then the variant at the centromeric HLA-DR locus followed by the alleles at C4A and C4B. Thus the common "supratype " 8/3-0-1 (the forerunner of what is now known as the "8.1" haplotype) has a clear set of allelic variants at HLA-A through to HLA-DR and such patients are predisposed to various generalised autoimmune diseases and insulin dependent juvenile onset diabetes mellitus.

About this time a curious issue arose which needs further attention. While many of the genes within the MHC deal with immune recognition and control of the self *versus* non-self discrimination in both innate and adaptive immunity, there are also many genes in this region that do not have an obvious immunological role. Why are these variant genes selected and preserved within the MHC region? Thus congenitial adrenal hyperplasia (CAH) due to 21-hydroxylase deficiency (21-OH°) associates with the 47/7-0-3 supratype; hereditary haemochromatosis (HH) is strongly associated with the common 7/15-3-1 supratype; and early onset Type 1 Insulin Dependent Diabetes Mellitus is strongly associated with the common 8/3-0-1 supratype. (All of these are now formally identified respectively as the 47.1, 7.1 and 8.1 ancestral haplotypes.)

Thus in the early 1980s the Perth group was concerned that the term "haplotype" had been widely misused. They recommended it's use needed to be restricted to "supratype" patterns that clearly segregated within an informative multigenerational family.

Further Reading Chapter 2 (and Addenda)

Alper, C.A., Raum, D., Karp, S., Awdeh, Z.L. and Yunis, E.J. (1983) Serum complement 'super-genes' of the major histocompatibility complex in man (complotypes), Vox Sang. 45 : 62-67.

Awdeh Z.L., Raum, D., Yunis, E.J. and Alper, C.A. (1983) Extended HLA/complement allele haplotypes: Evidence for T/t-like complex in man. Proc Natl. Acad, Sci. 80: 259-263.

Ceppellini, R., Curtoni, E.S., Marttiuz, P.L., Miggiano, V., Scudeller, G. and Serra, A. (1967) Genetics of leukocyte antigens: a family study of segregation and linkage. In Curtoni, E.S., Mattiuz, P.L., Tosi, R.M. (eds): Histocompatibility Testing 1967. Copenhagen, Munksgaard, 1967. Pp149-187.

Dawkins RL, Christiansen, F.T., Kay, P.H., Garlepp, M.J., McCluskey, J., Hollingsworth, P.N. and Zilko, P.J. (1983) Disease associations with complotypes, supratypes and haplotypes. Immunol. Rev. 70: 1-22.

Dawkins R.L., Christiansen, F.T. and Zilko, P.J. Editors (1982) "Immunogenetics in Rheumatology : Musculoskeletal Disease and D-Penicillamine". Excerpta Medica. Amsterdam-Oxford-Princeton.

McClusky J., Kay, P.H., Stuckey, M., Christiansen, F.T., Dawkins, R.L. and Wilson, G. (1983). MHC "Supratypes" Predicting Heterozygous 21-Hydroxylase Deficiency. The Lancet April 2, 764-765.

McCluskey, J., McCann, V. J., Kay, P. H., Zilko, P. J., Christiansen, F. T.. O'Neill, G. J. and Dawkins, R. L. (1982) HLA and Complement Allotypes in Type I (Insulin Dependent) Diabetes. Diabetologia 24: 162-165.

O'Neill, G. J., Nerl, C, Kay, P. H., Christiansen, F. T., McCluskey, J. & Dawkins, R. L. (1982) Complement C4 is a marker for adult rheumatoid arthritis. Lancet ii, 214-215

Raum, D., Awdeh, Z., Anderson, J., Strong, L., Granados, J., Teran, L., Giblett, E., Yunis, E.J., Alper, C.A. (1984) Human C4 haplotypes with duplicated C4A or C4B. Am. J. Hum. Genet. 36 : 72-79.

Rigby, R.J., Dawkins, R.L., Wetherall, J.D., and Hawkins, B.R. (1978) HLA in systemic lupus erythematosus : influence on severity. Tissue Antigens 12: 25-31.

Svejgaard, A and Ryder, L.P. (1977) : In HLA and Disease p 46, Editors: J. Dausset and A. Svejgaard, Munksgaard, Copenhagen.

Terasaki, P Editor (1990) History of HLA : Ten Recollections. UCLA Tissue Typing Laboratory, 1000 Veteran Avenue, Los Angeles, California, and The Regents of the University of California.

Uko G., Christiansen, F.T., Dawkins, R.L. and Kay, P.H. (1983) Low serum C4 concentrations in insulin dependent diabetes mellitus, Br. Med. J. 286 : 1748-1749.

****

Chapter 3

Discovery Of Ancestral Haplotypes

IRF/CYO research in the decade 1982-1992

Molecular Confirmation of Ancestral Haplotypes. International Collaborations - Homozygous HLA typed Cell Lines. Pulsed Field Gel Electrophoresis (PFGE). PFGE: Extensive Deletions and Insertions in Different MHC Supratypes - Tokunaga et al 1988. Discovery of the Haplospecific Geometric Element (HGE). Studies in lead up to discovery of the HGE. Use of Segmental Duplications to Detect Polymorphism. Ancestral Haplotypes: Conserved Population MHC Haplotypes. Further reading

Molecular Confirmation of Ancestral Haplotypes

By 1983 the serology used to define HLA markers was clearly showing associations of particular "supratypes" with particular sets of disease predispositions. By the mid to late 1980s the IRF/CYO group were confidently and reproducibly isolating long stretches of chromosomal fragments which contained all the genetic serologically markers defined by a particular "supratype". This allowed even more formal categorisation of what they became to term "Ancestral Haplotypes". All this was made possible by use of HLA typed homozygous B cell lines. And then, by 1992, the group made a very significant discovery which not only speeded up their work on the genomic structure of the MHC through the 1990s, but had huge implications for the successful matching of unrelated donors and recipients in bone marrow transplantation.

International Collaborations – Homozygous HLA typed Cell Lines

The role, during this period, of the International Histocompatibility Work Shops and Asia-Oceania Histocompatibility Workshops should not be underestimated. It was multicentre cooperation and collaboration on an international scale that Roger Dawkins (and others) lament the passing. It was genuine co-operation between all the main players. Standards of sera and typed "DNA" were maintained. The use of Epstein Barr Virus (EBV) - induced B lymphocyte cell lines from individuals who were homozygous across most of the ~ 4 Mb region of the MHC region allowed major cell repositories and typed collections to be set up. This made the cross checking of HLA typing data more effective and efficient.

During this period cell panels of EBV transformed - B cell lines with well defined MHC haplotypes were commonplace. Most of the useful lines were homozygotes, representing a large number of different ancestral haplotypes as the polymorphism detected in the MHC was invariably "haplotypic" and conserved *en bloc* and revealed in unrelated individuals. These panels accelerated research on ancestral haplotypes and allowed gene cloning and DNA sequencing of MHC regions from homozygotes.

Thus the concept of the "ancestral haplotype" - a long DNA sequence handed down unchanged from remote ancestors in the human-primate line - was being formally recognised by the multi-centre collaborations.

The HLA homozygous cell lines were of the utmost importance for the next step forward viz. long chromosomal structural analysis. Indeed the IRF/CYO group began to deploy the new techniques of molecular genetics and genomics using DNA isolated from HLA-typed homozygous cell lines by Pulsed field Gel Electrophoresis (PFGE). Applying new technology quickly has been a signal feature of the group in its quest to thoroughly investigate, and thus firmly establish, the molecular features of the ancestral haplotype concept.

Pulsed Field Gel Electrophoresis (PFGE)

The PFGE technique emerged in several laboratories in 1986-1988 and the IRF/CYO group applied it immediately to their problem of "supratypes" and disease associations, seeking explanations at the molecular chromosomal level.

The technique allowed the electrophoretic separation of very large genomic DNA fragments. The science of "Genomics" was born at this time within the MHC and via clinical necessity to improve transplantation procedures and the molecular data on disease associations (groups led by Hugh McDevitt, Jack Strominger, Roger Dawkins and others).

PFGE: Extensive deletions and insertions in different MHC supratypes - Tokunaga et al 1988.

The Perth group reasoned that since supratypes appeared to define entire chromosomal segments of at least 2 Mb or more, they felt confident enough to postulate that a given supratype would contain its own unique set of genetic features such as new genes and alleles, indels, copy number variations, duplications etc and that at least some of these supratype-specific patterns could be relevant to function and disease. This molecular prediction was proposed a year or so earlier at the 10th International Histocompatibility Work Shop in New York.

The laboratory of Ed Southern at University of California in San Diego (UCSD) had earlier developed the widely used "Southern Blotting" hybridization technique for identification of separated DNA fragments in standard agarose gel electrophoretic separations. The Southern group through Rakesh Anand, developed a particularly successful apparatus for PFGE. Accordingly, Anand was co-opted to the cause by Roger Dawkins and the group published their first important PFGE paper of a series in 1988 with Katsushi Tokunaga as the lead author.

Table 4 – Some Homozygous B cell Lines used by Tokunaga et al 1988

- HLA A3,Cw7, B7, C2C, BfS, C4A3, C4B1, DR2 (3,**7**,2)
- HLA A1Cw7, B8, C2C, BfS, C4AQ0,C4B1, DR3 (1,**8**,3)
- HLA B18, C2C, BfF1, C4A3, C4BQ0, DR3 (**18**,F1,3)
- HLA Bw65, C2C, BfS, C4A2, C4B1+2, DR1 (**65**, 1+2, 1)
- HLA Bw57, C2C, BfS, C4A6, C4B1, DR7 (**57**,6,7)

High molecular weigh DNA was carefully isolated to avoid shearing of DNA strands from EBV-transformed B cell lines from unrelated subjects, in most cases homozygous, for the five different MHC supratypes listed in Table 4. The bold number in the last bracket indicates the HLA-B allele and the future “ancestral haplotype” identification number the group would assign. Even at this advanced stage there was a reluctance in the Perth group to use the term "haplotype".

The high molecular weight DNA was gently broken up into defined fragments with rare restriction enzyme site cutters allowing the generation of large DNA fragments of the order of 2 Mb (The restriction enzymes *Not I*, *Pvu I* were the most useful in this regard.) Thus the very large DNA fragments of >500 Kb were separated over 35 hours PFGE runs (120V, 90s pulses) and smaller fragments (<500Kb) fragments separated over 65 hour runs (90V, 60s pulses). The gels were blotted to membranes and radioisotope labelled hybridisation probes specific for different genetic regions within the MHC used to reveal the markers carried by a particular defined restriction enzyme-generated DNA fragment. The labelled probes for MHC regions encompassed DP α, DQ α, DR α, DR β, 21OH, Bf, TNF, and HLA-B,-C class I genes. After exposure to X-ray film the membranes were re-probed with a different probe after washing and stripping.

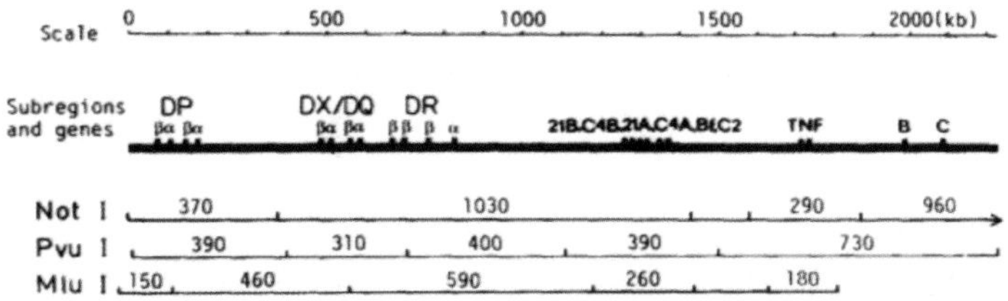

Figure 5 Genomic map of the human MHC in 3,7,2 supratype. Pulsed Field Gel Electrophoresis. Adapted from Fig. 2 from Tokunaga et al 1988 J Exp Med 168:933. This is the contiguous genomic map of the 3,7,2 supratype. Not I, Pvu I, Mlu I indicate sites where these restriction enzymes cut the DNA. The two Not I sites between class III and TNF genes were localized by using Not I + Pvu I double digests. No probe was available to determine sizes of the Not I and Mlu I fragments between the central Class II region and TNF. Sizing errors are ~ 10 kb for the fragments of < 500 kb, and 20 kb for > 600 kb.The MHC Class II region spans from 0 to about 800 Kb; the MHC Class II region spans from about 1000 Kb to 1600 Kb; and the MHC Class I regions takes in sequence from 1600 Kb to beyond 2000 Kb.

Shown in Fig 5 is the PFGE genomic map of the human MHC in the 3,7,2 supratype. This map was used as the standard reference for the other supratype profiles in this study. For each homozygous supratype DNA from three different unrelated individuals was examined (= six unrelated haplotypes) with the same hybridisation banding pattern. Thus a supratype reflects a "*fixed structure throughout more than 1 Mb of the HLA B to DQ regions*" (note that some supratypes 'fragment due to recombination' at the ends of the region within Class I and between DQ and DP).

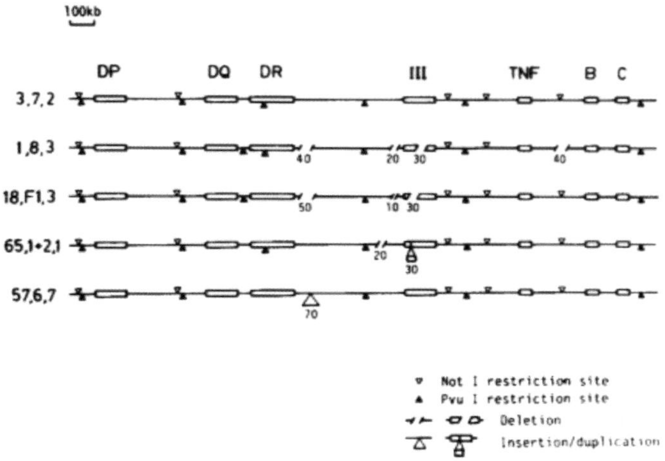

Figure 6 Genomic maps of the human MHC in different supratypes. Pulsed Field Gel Electrophoresis Adapted from Figure 3 Tokunaga et al 1988 J Exp Med 168:933.The figure shows the restriction maps with possible deletions, insertions/duplications specific for five different supratypes. The extent of the deletions and duplications is about 10 kb for most except 57,6,7 where it was about 20 kb.

The Perth group had achieved an important milestone - the complete molecular definition of a long extended region within the MHC suggestive of an ancestral haplotypic DNA sequence. The concept of *complete DNA sequence identity* in the Mb range begins to emerge - this was not a trivial conclusion.

As the five profiles were lined up it became apparent that there were extensive insertion/deletion and duplication events in comparison of DNA from different "Supratypes" - each had a characteristic molecular profile (Figure 6).

Tokunaga et al published the following summary of conclusions :

• It was very clear that different supratypes had characteristic deletions, insertions, duplications.

• The paternal and maternal regions within the MHC in an individual can be described as *en bloc* chromosomal segments (at least >1 Mb) and not a complex of separated genes.

• This implies that disease susceptibility associations as observed in the early 1980s with these supratypes are with *large chromosomal segments.* This further implies that all the polymorphic sequences in a segment were potentially involved in the "disease association."

• Accordingly then a "supratype" is a marker for a chromosomal segment that contains specific alleles at multiple loci over at least 1 Mb.

• Finally, since unrelated individuals with the same supratype have similar if not identical chromosomal segments of > 1 Mb this clearly "implies conservation of certain ancestral haplotypes."

These differences in segment length suggest that length could affect recombination frequency and provide a possible explanation for the lack of significant recombination within a polymorphic frozen block. While differences in supratype length could affect recombination frequency it cannot be the sole explanation for conservation. For example, one of the longest segments, the (**57**,6, 7) supratype, is common in both Caucasoids, Chinese and Negroids meaning it is a highly conserved ancestral haplotype (dating to before the out of Africa expansion).

The Perth group therefore reached another milestone and solid conclusion:

> " It seems likely that supratypes mark ancestral haplotypes that have some selective advantage conferred by particular combination of alleles at multiple loci .."

So we now have the clear emergence of the idea that the ancestral haplotype is the unit of inheritance rather than the single allele assumed under classical thinking. The issue of "selective advantage" of such ancestral haplotypes is still largely unresolved and is dealt with in a later chapter.

In a follow up paper the group used quantitative densitometry to measure the gene copy number carried by different MHC ancestral haplotypes after physical separation by pulsed field gel electrophoresis. For example the C4 and Cyp 21 genes in the central MHC can vary from copy numbers of one to three per ancestral haplotype. This method confirmed the gene dosage estimates by other methods, and provides a quantitative scale to gene expression effects on cellular gene function which affected a particular disease disposition.

To sum up this phase. We now have the concrete emergence of the important idea that the DNA sequence of ancestral haplotypes in unrelated individuals are *identical*.

I assume the group then reasoned that if they could analyse and characterize the polymorphisms in a short stretch of distinctive DNA sequence within a PFB in an ancestral haplotype sequence this could be considered to mark the entire PFB DNA sequence if not the entire ancestral haplotype. In my opinion the definitive data from the PFGE analysis not only confirmed the ancestral haplotype concept at the molecular level but it led the group on a technical search to simplify their molecular detection systems for AHs which led directly to their important technical discovery, the Haplospecific Geometric Element (HGE). This method uses a simple Polymerase Chain Reaction (PCR) assay that can detect long distinctive polymorphic frozen blocks rather than the more laborious PFGE method or even DNA sequencing (despite the fact that the latter is getting cheaper to perform by the day).

****

Discovery of Haplospecific Geometric Elements

For a number of years the discovery of haplospecific geometric elements (HGEs) has both surprised and challenged me to understand the molecular mechanism of the genesis of such elements. How can a relatively *short* HGE (< 1000bp) polymorphic sequence mark an entire haplospecific polymorphic DNA sequence of length ≥ 300-500Kb to 1Mb, if not an entire full length ancestral haplotype (≥3 Mb)? Indeed this feature of HGEs still surprises members of the Perth group (see McLure et al 2013). We will leave this problem until later. Here we are concerned with the key studies which led to their discovery.

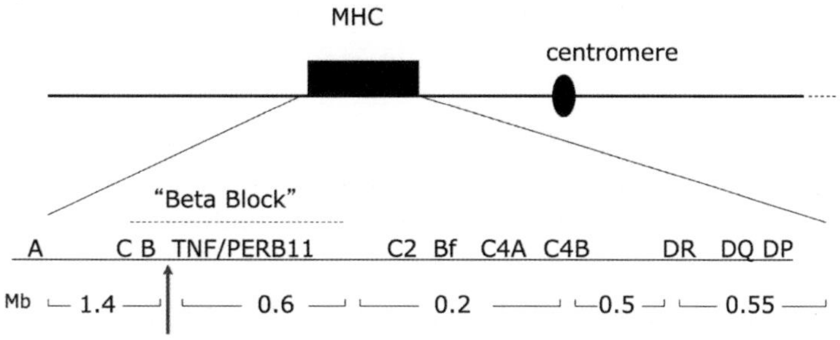

Figure 7 Haplospecific Restriction Fragment Length Polymorphism between HLAB and Tumour Necrosis Factor. Strategy adapted from Xu X et al 1992 Human Immunol 33: 89-97. Note PERB11 is now designated MIC.

Studies in lead up to discovery of the HGE

The precursor work by Xiongwen Wu and colleagues concerned investigations of the polymorphic sequences which existed between HLA-B and the Tumor Necrosis Factor gene (TNF) (Figs 7 and 8). They used Restriction Fragment Length Polymorphism (RFLP) analysis whereby the region was fragmented using preselected restriction enzymes to generate DNA fragments of known expected size (1 - 6 Kb) which then allowed conventional electrophoretic separation on agarose gels and characterisation of sequence content with various Southern blot hybridisation probes for that MHC region. Thus the fragments of DNA are far smaller that those generated for separation by PFGE in the earlier studies. Nevertheless, these DNA fragment separation analyses and the interpretation of gel banding patterns, and their identification in given haplotypic sequences, added to the molecular precision and expertise the Perth group was continuing to develop.

A large panel of EBV-induced B cell lines containing 37 homozygotes representing 21 different ancestral haplotypes was studied (Table 5). The hybridisation probes (Fig 8) were provided by the group of Jack Strominger at Harvard Medical School.

A typical set of allelic banding patterns generated by *Taq1* and *RSA1* restriction enzymes with the X, V and Y probes (Figs 9-11) is shown summarised in Table 6. The X probe following cutting with *Taq1* revealed three bands tested across the DNA panel, L (5.2Kb), M (4.5 Kb) and S (3.3 Kb). The V probe following *Taq1*

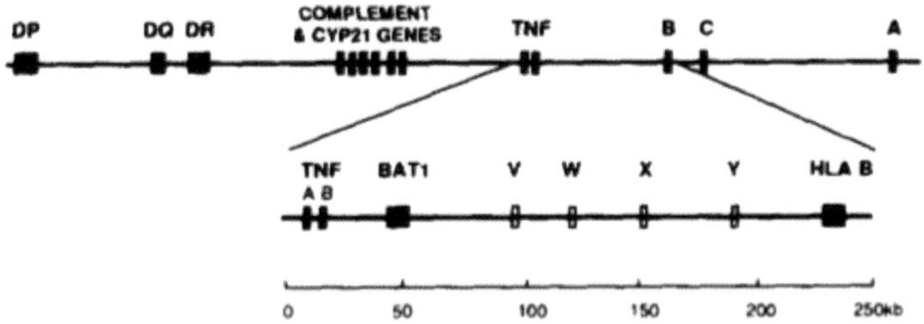

Figure 8 The approximate location of genomic hybridization probes Y,X,W,V between HLA-B and TNF. Adapted from Fig 1 from Xu X et al 1992 Human Immunol 33: 89-97. The HLA Class I region is on the right and the HLA Class II region on the left.

digestion generated six fragment sizes (0.88Kb, 1Kb, 1.05Kb, 1,9Kb, 2.2 Kb and 3,1Kb) which fell into distinct allelic patterns A, B, C, D. When both enzymes digested the DNA the Y probe detected six different band sizes (1.69Kb, 1.73Kb, 1.77Kb, 1.80Kb, 1.85Kb, 1.88Kb) with seven different allelic patterns A through G (Table 6) .

They found that the region between HLA B and TNF loci is highly polymorphic and the Y-X-V haplotypes segregate with HLA haplotypes. Thus homozygous

AH	HLA A	Cw	B	C2	Bf	C4A	C4B	DR	DQw	Race
7.1	3	7	7	C	S	3	1	15	6	C
7.2	24	7	7	C	S	3 + 3	1	1	5	M
8.1	1	7	8	C	S	Q0	1	3	2	C
13.1	30	6	13	C	S	3	1	7	2	C/M
18.1	25	?	18	Q0	S	4	2	15	6	C
18.2	30	5	18	C	F1	3	Q0	3	2	C
35.1	?	4	35	?	S	3	1	11	7	C
42.1	?	2	42	C	F	1	Q0	3	4	N
44.1	2	5	44	C	S	3	Q0	4	7	C
44.2	29	4	44	C	F	3	1	7	2	C
44.3	29	?	44	?	S	Q0	1	7	2	M
46.1	2	11	46	C	S	4	2	9	9	M
46.2	2	11	46	C	S	4	2	8	6	M
47.1	3	6	47	C	F	91	Q0	7	2	C
52.1	24	?	52	C	S	3 + 2	Q0	15	6	M
54.1	?	1	54	C	S	3	5	4	4	M
57.1	1	6	57	C	S	6	1	7	9	C/M/N
60.3	2	3	60	C	S	Q0	2	13	6	C
62.1	2	3	62	C	S	3	3	4	8	C
62.2	?	3	62	B	S	4	2	4	8	C
65.1	?	8	65	C	S	2	1 + 2	1	5	C

Table 5 – Ancestral haplotypes used in study by Wu et al 1988 Hum Immunol 33 :89-97 . Adapted from Table 1. A minimum of at least 2 different homozygous cell lines were use to study each ancestral haplotype. ?, not detected; C= Caucasoid; M= Mongoloid; N = Negroid.

cells for AHs provide allelic pattern predictions for heterozygous cells with the Y-X-V haplotypes being haplotypic and often haplospecific. Y, X and V allelic patterns are inherited within a unit family and segregate with HLA haplotypes (Fig.11). An example of the banding patterns for probe X on *Taq1* digested DNA across all ancestral haplotypes examined is shown in Figure 9. Note the three bands L, M, and S and their patterns across the homozygous DNAs. As expected, more complex patterns are revealed using the V probe on *Taq1* digested samples (Figure 10). And it is clear these are stable haplospecific patterns as the Y, X, and V patterns are stably inherited within a nuclear family (Fig 11).

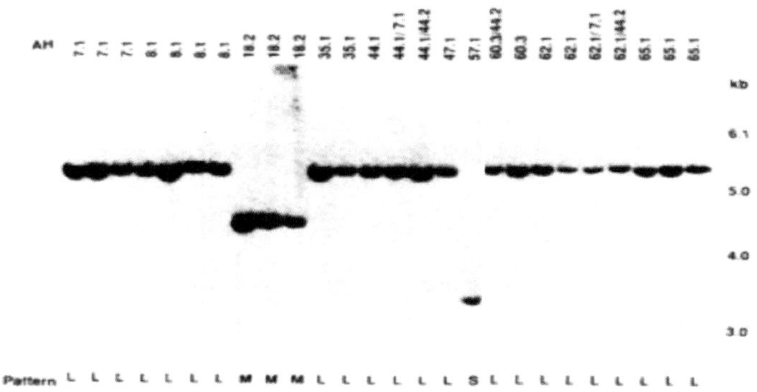

Figure 9 Three different fragments may be seen after Taq 1 digestion and probing for X. Adapted from Fig 2 from Xu X et al 1992 Human Immunol 33: 89-97. A sample of the data only is shown. Three different fragments seen after Taq 1 restriction enzyme digestion and probing for X (see Fig 8). Each ancestral haplotype carried by the homozygous cell line yields a single fragment and each can be assigned to each haplotype. The molecular weight ladder (kb) is on the right.

All the banding patterns are summarised in Table 7. It is clear that ancestral haplotypes sharing the *same* Y-X-V haplotypes show extensive similarities elsewhere within the MHC. We witness here again the emergence of the concept that a shorter defined region of DNA (the probe) may mark a much larger DNA sequence in a haplospecific manner within the MHC.

Thus by the late 1980s early 1990s the Perth group had now become experts at analysing haplotypic and haplospecific banding patterns following agarose gel electrophoresis. These technical skills meant that the next step for HGE development was a relatively small one.

Table 6 Designation of allelic patterns detected by genomic probes X,V and Y

Probe	Enzyme	Pattern	Fragments (kb)
X (R17A)	Taq 1	L	5.2
		M	4.5
		S	3.3
V (R9A)	Taq 1	A	0.88, 1.9. 3.1
		B	0.88, 1.05, 1.9. 2.2
		C	0.88, 1.00, 1.9, 2.2
		D	0.88, 1.05, 1.9, 3.1
Y (M20A)	Taq 1 + Rsa 1	A	1.69, 1.88
		B	1.73, 1.88
		C	1.69, 1.80
		D	1.77, 1.88
		E	1.77, 1.85
		F	1.69,1.73
		G	1.69, 1.85

Table 6 Designation of allelic patterns detected by genomic probes X,V and Y. Adapted from Table **2** Xu X et al 1992 Human Immunol 33: 89-97. Allelic patterns (RFLP pattern) were designated to various combinations of fragment sizes detected by probes, X, V and Y as shown, digested with either Taq 1 fro X and V probing, and with Taq 1 and Rsa 1 for Y probing (see Fig 8). Note the three different patterns for X, the 4 distinct patterns for V and the 7 different combinations of fragments sizes for Y.

AH	A	Cw	B	Y	X	V	C2	Bf	C4A	C4B	DRB1	DQB1
52.1	24	?	52	A	L	A	C	S	3+2	0	1502	0601
7.1	3	7	7	A	L	A	C	S	3	1	1501	0602
7.2	24	7	7	A	L	A	C	S	3+3	1	0101	0501
46.1	2	11	46	A	L	C	C	S	4	2	0901	0303
46.2	2	11	46	A	L	C	C	S	4	2	0803b	0601
44.1	2	5	44	A	L	C	C	S	3	0	0401	0301
62.1	2	3	62	A	L	B	C	S	3	3	0401	0302
62.2	?	3	62	A	L	B	B	S	4	2	0401	0302
60.3	2	3	60	A	L	B	C	S	0	2	1302	0604
13.1	30	6	13	D	L	C	C	S	3	1	07	0201
44.2	29	4	44	D	L	C	C	F	3	1	07	0201

Table 7 Extensive similarities among the AHs sharing Y-X-V haplotypes. Adapted from Table 5 in Xu X et al 1992 Human Immunol 33: 89-97. HLA A, Cw, B and complement alleles were determined by serology. DRB1 and DQB1 were based on SSO typing. The boxed areas encompass regions of similarity between AHs. Two to eight different AH bearing cell lines examined, each cell line designated with the same MHC AH gave the same pattern.

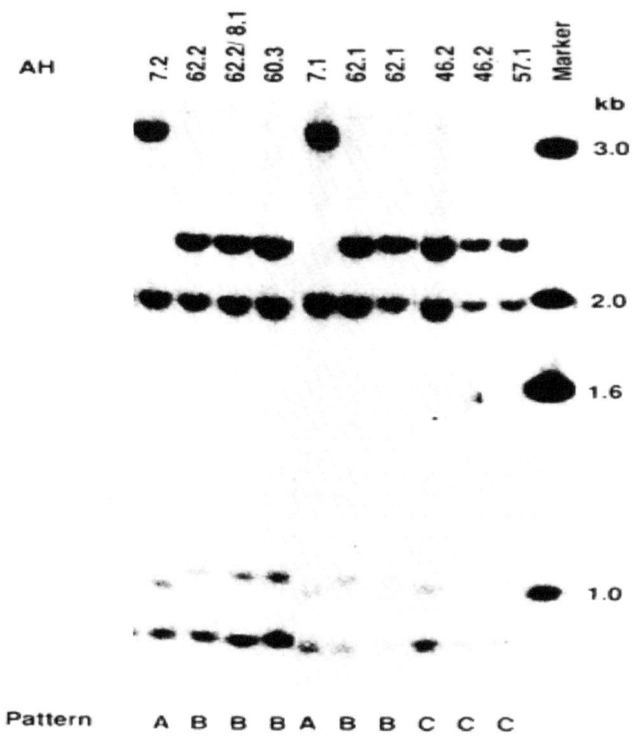

Figure 10 After Taq 1 digestion and probing with probe V six fragments were seen but two appear to be nonpolymorphic. Adapted from Figure 3 from Xu X et al 1992 Human Immunol 33: 89-97. Each polymorphic cell line designated with the ancestral haplotype (AH) carries two of the four polymorphic fragments. Patterns designated A, B, C.

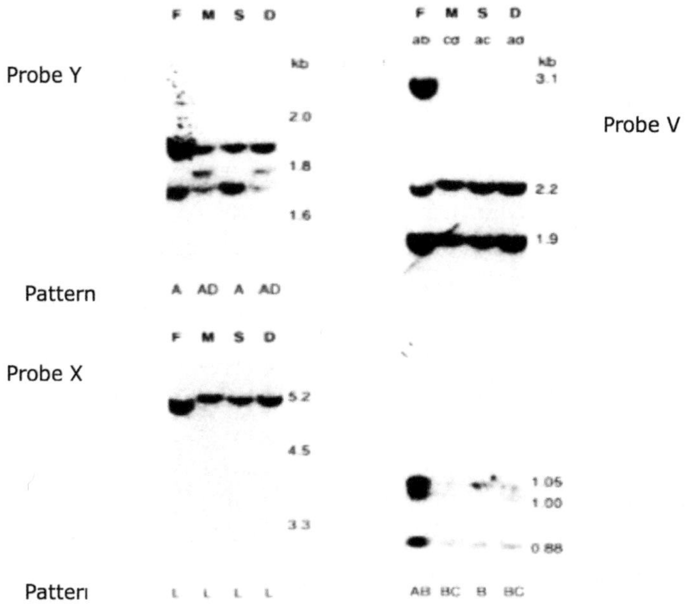

Figure 11 Y, X and V allelic patterns are inherited within family DAR members and segregate with HLA haplotypes. Adapted from Figure 5 from Xu X et al 1992 Human Immunol 33: 89-97. Y, X and V allelic patterns are inherited within family DAR members and segregate with HLS haplotypes. (See original figure for complete HLA haplotype patterns). The Y-X-V haplotypes segregated with HLA haplotypes and the genotypesare in agreement with those derived from unrelated homozygous subjects carrying the same AHs. F= father, M= mother, S= son, D=daughter.

Use of Segmental Duplications to Detect Polymorphism

A critical strategic step was taken by Roger Dawkins in the late 1980s and certainly by 1991. The decision was made to analyse MHC polymorphisms between DNA sequences of *duplicated regions*. In personal interview Dawkins recalls a discussion which triggered this line of thinking at an International Histocompatibility Meeting in the late 1980s. He was in the lobby of the Waldorf Astoria in New York involved in a discussion with John Trowsdale and Duncan Campbell on the significance of repetitive retroelements and retropseudogenes within the MHC, and whether they should be considered "junk" DNA or not. When questioned further by the author on these conversations Roger Dawkins recalls saying to Duncan Campbell " Have you noticed that all the polymorphic products arise from duplicated genes." These discussions were significant and encouraged the decision to analyse the sequence divergence of duplicated regions – this turned out to be another *key step* in the genomic analysis of the MHC by the Perth group.

It was the great Japanese scientist and scholar Susumo Ohno who wrote the definitive book on gene duplication. He considered it as one of the main building processes of the evolution of genomes and their diversity (particularly tandem duplication via unequal crossing over, e.g. Fig 12). Thus the duplicated gene not only increased chromosome length but the duplicate gene was free to mutate while the original copy held its function. There have been criticisms of this simple schema as there could be quantitative consequences of gene dosage effects which may not be neutral as discussed earlier. In any case, the analysis of polymorphisms of tandem segmental duplications was a key step forward by the Perth group. The schemes in Fig 12a,b follow the Retroviral Induction Model for segmental duplication and ancestral haplotype generation discussed in a later chapter.

Fig 12a Unequal Crossing Over - Meiotic Duplication/Deletion

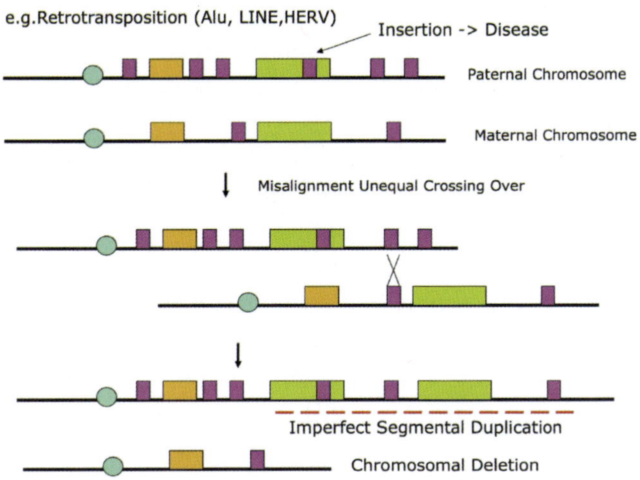

Fig 12b Unequal Crossing Over - Meiotic Duplication/Deletion

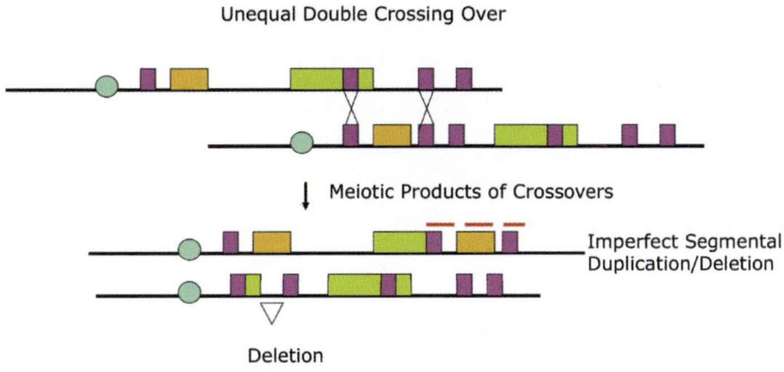

Lawrence Abraham and colleagues selected a duplicated region between HLA-B and TNF (Figure 13) to analyse differences between different MHC ancestral haplotypes in the beta block. They asked the question " Does the degree of polymorphism found within expressed protein products of HLA-B and HLA-C correlate with surrounding polymorphism at the genomic level? "

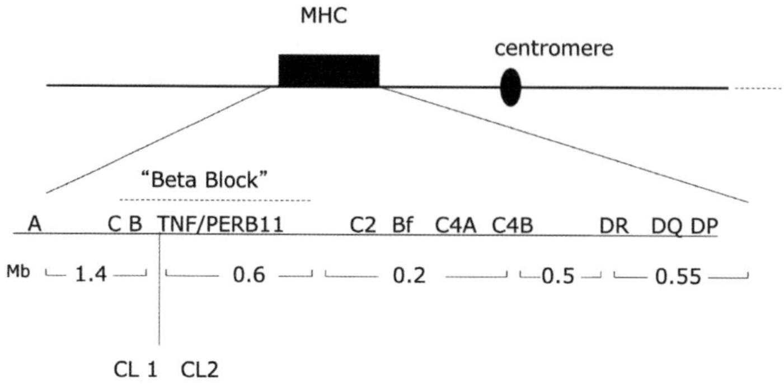

Figure 13 Strategy using CL1 and CL2 PCR primers to amplify haplospecfic geometric elements (HGE) within the beta block of the MHC. Abraham et al 1992 Tissue Antigens 39: 117-121

The next step in the line of reasoning is so important I have decided to reproduce it in full with the scientific jargon of the time in association with the MHC context in Figure 13:

> " "Using genomic walking probes ..... in the first instance, we walked in the region between B144 and HLA-B until we found a sub-region where there were major differences between the four cell lines (Leelayuwat et al, unpublished). Also, after probing DNA from a panel of homozygous cell lines, we were able to demonstrate extensive polymorphism at the RFLP level ..... Further more, there was evidence that parts of the sub-region are duplicated with an interval of some 20-35 Kb between homologous sequences designated CL1 (telomeric) and CL2 (centromeric)."
>
> "Sequence analysis of selected lambda clones showed that much of the polymorphism detected at the RFLP level is accounted for by in sertions/deletions of sequences which we refer to provisionally as haplospecific geometric elements ..... Although there are features which suggest similarity to "satellites", we are most impressed

> by the geometry (relations of magnitudes) when different haplotypes are compared, ie. the lengths of the elements vary apparently because certain stretches of sequence are inserted or deleted (Fig 1). The sequence contains (but is not necessarily restricted to) reiterated di-nucleotides and possibly other non-random patterns. There is also some apparent symmetry around the region which is deleted or inserted (Fig. 1). " (Abraham et al 1992).

Please note that "satellites" in this context refers to "microsatellites" which are simple repetitive di-, or tri-nucleotide repeats often generated by polymerase "slippage" errors during DNA replication of such regions, leading to "micro-satellite " expansions. The authors, in this context, are emphasizing that the present features are *quite different*. Indeed as they later point out such HGEs

> .. "ideally contain imperfect repetitive units that exhibit some additional forms of geometric complexity. The rationale is that complex, imperfect elements are less prone to slippage and mutation, and therefore more stable than simple, perfect repeats such as di-nucleotide microsatellites. Since the aim is to identify polymorphism indicative of AHs, stability throughout human evolution is essential."(McLure et al 2013).

The first clear haplospecific patterns of PCR amplification products in this region of the MHC, and their associated ancestral haplotypes in which they appear, are displayed in the DNA sequence data of the polymorphic PCR products in Table 8 and some of the agarose gel patterns of the various PCR product sizes in Figure 14. It can be seen that three unrelated homozygotes (= 6 chromosomes) carrying the ancestral haplotypes 46.1 all have an *identical banding pattern* which in turn is different from the specific haplospecific banding patterns in homozygous DNA from ancestral haplotypes 18.2, 8.1 and 7.2

These are very striking data. They underline the point foreshadowed above - a very long segment of DNA defining an ancestral haplotype *can be marked specifically* by the pattern of indels flanked by the closely spaced conserved PCR priming sites! As indicated these are surprising findings.

				number of variable nucleotides
57.1 - CL1	T C A G A	$(TC)_{12}(TG)_{6}(TC)_{14}(TG)_{3}(TC)_{12}$	T G T T T	94
18.2 - CL1	. . . . .	$(TC)_{14}$	. . . . .	28
8.1 - CL1	. . . . .	$(TC)_{28}$	. . . . .	56
7.1 - CL1	. . . . .	$(TC)_{12}(TG)_{5}(TC)_{14}(TG)_{3}(TC)_{12}$	. . . . .	94

Table 8 First HGE patterns. Adapted from Fig 1 Abraham et al 1992 Tissue Antigens 39: 117-121. Genomic elements at CL1 are polymorphic and ancestral haplotype-specific. The nucleotide sequences of the geometric elements and flanking regions at CL1 for the 57.1, 8.1 and 7.1 Ancestral haplotypes are shown. The CL1 locus from 18.2 is also presented. The number of nucleotides comprising each geometric element is listed. - , consensus sequences, ( )n = number of dinucleotide repeats.

As we will discuss shortly, HGEs were then used as the basis of the Genomic Matching Technique (GMT) to predict successful outcomes in organ/tissue transplantation between unrelated donor and recipients. The technique was successfully used through the 1990s and 2000s to detect and analyse PFBs on other human chromosomes and duplicated DNA regions in other mammalian species. Indeed Roger Dawkins and his IRF/CYO team have become experts in selecting potential informative HGEs (revealing polymorphism) in duplicated genomic regions assisted by the “Dot Plotter” bioinformatic tool. Their success, particularly in the "duplotyping" survey across the human genome at 109 duplicated genomic sites, means that most later prospecting trials would have been successful in selecting those HGEs very likely to detect polymorphism at that duplicated site (McLure et al 2013).

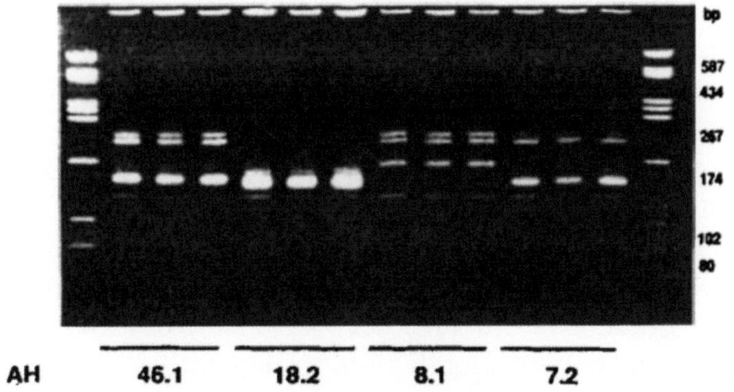

Figure 14 First Gel Patterns using HGE Adapted from Figure 2 Abraham et al 1992 Tissue Antigens 39: 117-121. DNA was extracted from 12 homozygous cell lines representing 3 examples each of ancestral haplotypes (AH) : 46.1, 18.2, 8.1, 7.2, and PCR amplification proceeded using primers flanking the geometric elements. Note the reproducible haplotypic patterns for all four ancestral haplotypes after electrophoresis in the agarose gel. The molecular weight markers are in the extreme left and extreme right lanes.

By this it is not implied that such surveys are complete, as there are likely to be regions of important polymorphic frozen blocks not picked up in the initial McLure et al survey.

To repeat, later in this book we will discuss some possible mechanisms for how a distinctive HGE is laid down. Such sites, mainly in the intergenic regions, exist in the context of a long contiguous segment of DNA, with its associated haplotypic profile of indels, duplications, SNPs and retroelements that defines that polymorphic block region of an ancestral haplotype.

****

Ancestral Haplotypes – Conserved Population MHC Haplotypes

Clearly 1992 was a big year for the Perth group. Not only did they publish their data on the discovery of HGEs but in that year they published a valuable paper by Mariapia Degli-Esposti and her colleagues which formalised the concept of MHC ancestral haplotypes inherited from remote ancestors. This study also gave a formal estimate of the MHC diversity represented by ancestral haplotypes and their recombinants in a well studied racial group, viz. in the Caucasoid population.

In 1992 Degli-Esposti and her colleagues reviewed their data from the previous decade on Caucasoids and asked some important questions among which were:

1. Of the supratypes previously described, and their disease associations, which of them can be identified with an ancestral haplotype - that is by the criterion of their appearance in unrelated individuals, and then by stable inheritance (segregation) in a nuclear, or better still, three generation family?

2. What are the characteristics of "rare supratypes" - are they recombinants of existing ancestral haplotypes?

3. For the key Class I HLA-B alleles which ones are shared by more than a single ancestral haplotype?

4. Amongst the haplotypes in unrelated individuals what proportion can be accounted for by ancestral haplotypes?

The answer to the last question is critical in understanding inherited MHC diversity. Recall that the definition of a haplotype is arbitrarily dependent on its observed frequency in the study population. For inclusion a cut off is often set at a frequency of $\geq 1\%$ and therefore the larger the study population the better.

They therefore needed a large number of supratypes that they could assign unequivocally to haplotypes in transmission through at least two generations of a nuclear family. The further identification criterion was that a supratype had to be carried by three or more unrelated nuclear haplotypes (≥ 3 different families). Thus alleles were only assigned to a haplotype where segregation could be unequivocally demonstrated. For those supratypes that could not be analysed by family studies they chose those occurring three or more times to see if their pattern could be explained by historical recombination between two established ancestral haplotypes. The rare supratypes which could not be assigned or explained by one historical recombination event were left out of the analysis - they could be complex multiple recombinants or just rare combinations arising from past racial admixture.

Their data included all haplotypes assigned from family studies in the Department of Clinical Immunology at Royal Perth Hospital in the period 1980 to 1990. The complete database included 836 haplotypes - seven were then excluded prior to analysis because of possible recombination within a nuclear family. This left 829 haplotypes from 219 unrelated families in what they called "data set 1."They then created a subset of data set 1 by:

> "excluding all records where reason for study was the presence of an HLA-associated disease (ankylosing spondylitis, arthritis, complement deficiency, hemochromatosis, insullin-dependent diabetes mellitus (IDDM), IgA deficiency, myastheniagravis, myositis, nephritis, rheumatoid arthritis, Sjogren's syndrome, systemic lupus erythematosus, or congenital adrenal hyperplasia) or where the reason for study was unknown."

This was referred to as "data set 2" and included 348 haplotypes from 94 families - and these families were studied for the following reasons:

" bone marrow transplant, renal transplant, C3B deficiency, Ewing sarcoma, Hodgkin's disease, immune deficiency (not complement or IgA deficiency), melanoma, cryoglobulinemia, thalassemia major, or Wilson's disease."

In Table 9 is the list of the 23 supratypes and disease associations published by the Perth group 1980-1990. The data analysis in sets 1 and 2 proceeded as follows; haplotypes were sorted according to the HLA-B allele (that would designate the haplotype identification

HLA-			Central non-HLA				HLA-			
A	Cw	B	C2	Bf	C4A	C4B	DR	Dw	DQw	Disease associations
3	7	7	C	S	3	1	15	2	6	MS, CD, SLE, Hemochr
1	7	8	C	S	Q0	1	3	3	2	GMG, IDDM, SLE
30	6	13	C	S	3	1	7	?	2	RT
25	?	18	Q0	S	4	2	15	2	6	C2 Def, SLE
30	5	18	C	F1	3	Q0	3	3	2	IDDM
?	?	18	?	S	3	1	11	?	7	RT
?	4	35	?	S	3	1	11	?	7	RT
3	4	35	C	F	3 + 2	Q0	1	1	5	HIV rapid progression
11	4	35	?	S	3	Q0	1	1	5	HIV rapid progression
2	5	44	C	S	3	Q0	4	4	7	Felry's RA
29	4	44	C	F	3	1	7	17	2	CD, IgA Def
3	6	47	C	F	1	Q0	7	?	2	CAH
?	3	55	B	S	4	5	14	?	?	RT
1	6	57	C	S	6	1	7	11	9	IgA Def, Psoriasis
?	3	60	C	S	3	1	4	?	3	IDDM
?	3	60	C	S	3	Q0	8	8	4	RT
2	3	60	C	S	Q0	2	13	19	6	RT
2	3	62	C	S	3	3	4	4	8	IDDM, RA
?	3	62	B	S	4	2	4	4	8	IDDM, RA
?	?	62	C	F	3	1	13	?	6	RT
?	8	64	C	S	3	1	7	7	2	RT
?	8	65	C	S	2	1 + 2	1	20	5	IgA Def, 21OH Def
?	8	65	C	F	3	1	13	19	6	RT

Table 9 List of Caucasoid supratypes and their disease associations. Adapted From Table 1 Degli-Esposti et al 1992 Hum Immunol 34:242-252. Note that Supratypes are listed in numerical order according to their HLA-B allele. CAH, congenital hemochromatosis; CD, celiac disease; Def, deficiency; GMG, generalised myasthenia gravis; IDDM, insulindependentdiabetesmellitus; MS, multiplesclerosis; Hemochr, hemochromatosis; RA, rhematoidarthritis; SLE,systemic lupus erythematosus; and RT, renal transplant survival .

number) then through the complement region Bf, C4A, C4B and then HLA-DR.

There are 49 supratypes carried on three or more unrelated haplotypes (Table 10). Two supratypes listed in

Table 9, B47, BfF, C4A1, C4BQ0, DR7 ; B62, BfF, C4A3, C4B1,DR13 occurred less than three times in data set 1.

The data in Table 10 were then further analysed to see if they could be explained by a single historical recombination between two well established ancestral haplotypes. This is shown in Table 11. In the left hand column are the list of 22 ancestral haplotypes the group was very confident of assigning. In the next column are nine new potential ancestral haplotypes uncovered that could be assigned on the assembled data but require more detailed genomic analysis using the group's molecular criteria at the DNA level - they appear likely to be rare Caucasoid ancestral haplotypes. In the next column on the right are haplotypes that clearly require more study. They are rare combinations that share some allele with other ancestral haplotypes - all of them carry the BfS, C4A3, C4B1 complotype.

HLA-B	Bf	C4A	C4B	HLA-DR	n
5	S	3	1	2	7
51	S	3	1	5	8
51	S	3	1	7	5
7	S	3	1	1	3
7	S	3	1	2	53
7	S	3	1	4	8
7	S	3	1	6	7
8	S	Q0	1	3	104
13	S	3	1	7	10
14	S	2	1 + 2	1	10
14	S	2	1 + 2	5	4
14	F	3	1	6	6
14	S	3	1	7	5
18	S	4	2	2	7
18	F1	3	Q0	3	11
18	S	3	1	5	7
21	F	3	2	4	3
21	S1	2	1 + 1.5	7	4
49	S	3	1	5	4
22	S	4	5	6	8
27	S	3	1	1	5
27	S	4	2	1	3
27	S	3	1	2	3
27	S	3	1	4	7

HLA-B	Bf	C4A	C4B	HLA-DR	n
35	F	3 + 2	Q0	1	6
35	S	3	Q0	1	8
35	S	3	1	4	3
35	S	3	1	5	7
35	S	3	1	6	5
37	F	3	1	6	3
39	F	2	Q0	1	3
39	S	3	1	2	4
40	S	3	1	4	10
40	S	3	1	6	3
60	S	Q0	2	6	9
60	S	3	Q0	8	3
44	S	3	1	2	7
44	S	3	Q0	4	26
44	S	3	1	4	4
44	S	3	1	5	12
44	F	3	1	7	19
44	F	Q0	1	7	3
57	S	6	1	6	3
57	S	6	1	7	20
62	S	3	3	4	14
62	S	3	1	4	5
62	S	3	2	4	3
62	S	4	2	4	4
62	S	3	1	5	3

Table 10 List of supratypes (occurring three or more times on a haplotype) sorted by their HLA-B allele. Adapted from Table 2 Degli-Esposti et al 1992 Hum Immunol 34:242-252

In the fourth column on the right (Table 11) are those haplotypes that can be accounted for by a single recombination involving one of the haplotypes in the first column. It should be noted from Table 11 that intact ancestral haplotypes occur *more frequently* than their putative recombinants, attesting to the inherited stability of the higher frequency ancestral haplotypes.

Shown in Table 12 is the estimate of the frequency of ancestral haplotypes and their recombinants in the local (Perth area, South West WA) Caucasoid population using the 348 haplotypes from 94 unrelated families of data set 2 ; that is, those studied for reasons other than the presence of a HLA-associated disease (to avoid inflating the estimate of ancestral haplotypes in the population).

Over 73% of the population haplotypes carried "at least 400Kb (HLA-B-complotype=600Kb; and complotype-DR=400Kb) " of a known ancestral haplotype. The 27% remaining could include haplotypes with shorter fragments (see Yunis et al 2003) and are possibly rare ancestral haplotypes, now postulated to account for rare alleles not carried on the current list of 22.

This milestone study arrived at a very important conclusion which in the author's opinion is not widely realised or appreciated. Perhaps greater than 70% of all Caucasoid MHC diversity can be accounted for by about 22 common Ancestral Haplotypes and their recombinant blocks thereof. Given that the definition of an ancestral haplotype depends on its population frequency, this is likely to exceed 90% if the arbitrary cut-off criterion for an AH is dropped from a population frequency of 1% to 0.1%-05%. In summary the authors conclude for their Caucasoid population:

> "Thus, the basic pool of polymorphism within the MHC can probably be described in terms of a relatively small number (perhaps 50-200) of ancestral haplotypes and most individuals will be found to possess at least parts of one or more of these ancestral haplotypes." (Degli-Esposti et al 1992.)

This leads to another important conclusion - all MHC SNP alleles and MHC marker alleles that may be discovered in the future in Caucasoids are likely to exist *as part of an existing ancestral haplotype.* It was another

Ancestral haplotypes						Possible ancestral haplotypes					Probable recombinants of ancestral haplotypes					Donor
HLA-B	Bf	C4A	C4B	HLA-DR	AH	HLA-B	Bf	C4A	C4B	HLA-DR	HLA-B	Bf	C4A	C4B	HLA-DR	AH
7	S	3	1	2	7.1	21	F	3	2	4	14	S	2	1 + 2	5	65.1
8	S	Q0	1	3	8.1	49	S	3	1	5	57	S	6	1	6	57.1
13	S	3	1	7	13.1	27	S	4	2	1						
18	S	4	2	2	18.1	27	S	3	1	4						
18	F1	3	Q0	3	18.2	37	F	3	1	6	7	S	3	1	1	7.1
18	S	3	1	5	18.3	39	F	2	Q0	1	7	S	3	1	4	7.1
35	S	3	1	5	35.1	44	F	Q0	1	7	7	S	3	1	6	7.1
35	F	3 + 2	Q0	1	35.2	51	S	3	1	5	5	S	3	1	2	7.1
35	S	3	Q0	1	35.3	62	S	3	2	4	27	S	3	1	2	7.1
44	S	3	Q0	4	44.1						39	S	3	1	2	7.1
44	F	3	1	7	44.2						44	S	3	1	2	7.1
50	S1	2	1 + 1.5	7	50.1											
55	S	4	5	6	55.1											
57	S	6	1	7	57.1											
60	S	3	1	4	60.1											
60	S	3	Q0	8	60.2											
60	S	Q0	2	6	60.3											
62	S	3	3	4	62.1											
62	S	4	2	4	62.2											
64	S	3	1	7	64.1											
65	S	2	1 + 2	1	65.1											
65	F	3	1	6	65.2											

[solid box] Well-defined ancestral haplotypes: the gross and detailed DNA structure of these haplotypes has been studied on three or more unrelated haplotypes.
[dashed box] The portion of a haplotype that may be explained by recombination with a well-defined ancestral haplotype.

Ancestral haplotype	HLA-B	Bf	C4A	C4B	HLA-DR	B–C4	B–DR	C4–DR	Total	
7.1	7	S	3	1	2	13	19	13	45	
8.1	8	S	0	1	3	1	40	5	46	
13.1	13	S	3	1	7	0	5	4	9	
18.1	18	S	4	2	2	0	3	1	4	
18.2	18	F1	3	0	3	0	4	2	6	
18.3	18	S	3	1	5	2	4	12	18	
35.1	35	S	3	1	5	5	5	14	24	
35.2	35	F	3 + 2	0	1	0	3	0	3	
35.3	35	S	3	0	1	1	5	2	8	
44.1	44	S	3	0	4	4	10	5	19	
44.2	44	F	3	1	7	2	6	1	9	
50.1	50	S1	1	1 + 1.5	7	1	1	0	2	
55.1	55	S	4	5	6	1	1	1	3	
57.1	57	S	6	1	7	2	6	1	9	
60.1	60	S	3	1	4	4	2	9	15	
60.3	60	S	0	2	6	1	6	2	9	
62.1	62	S	3	3	4	1	3	0	4	
62.2	62	S	4	2	4	2	2	4	8	
64.1	64	S	3	1	7	1	2	2	5	
65.1	65	S	2	1 + 2	1	0	2	0	2	
65.2	65	F	3	1	6	0	4	3	7	
									255/348	73.0%

Table 12 Frequency of ancestral haplotypes and their "simple" recombinants in an Australian population (Caucasoids) . Adapted from Table 5. Degli-Esposti et al 1992 Hum Immunol 34: 242-252

decade and a half for extended DNA resequencing of MHC haplotypes from the Seattle group of Dan Geraghty and John Hansen, to confirm the complete DNA sequence identity of unrelated MHC regions carrying the same ancestral haplotype (e.g. see Smith et al 2006). Whilst the Perth group conceded that other ways of describing ancestral haplotypes are possible they state:

> " .. we suspect that our conserved haplotypes are the same at the sequence level as the extended haplotypes of Alper and colleagues."
> (Degli-Esposti et al 1992.)

Table 11 Haplotypes occurring three or more times can be subclassified. Adapted from Table 3 Degli-Esposti et al 1992 Hum Immunol 34: 242-252. Solid boxes, well defined ancestral haplotypes; hatched boxes, the portion of a haplotype that may be explained by recombination with a well defined ancestral haplotype.

In follow up studies the group began to extend "ancestral haplotyping" to other ethnic and racial groups as well as into "recombinant mapping". Thus approaches were developed to localize susceptibility genes for various autoimmune and other diseases by comparative mapping with ancestral haplotypes and their recombinant ancestral haplotypes. This type of expanded study needs to be taken up and extended by other laboratories on a wider scale.

The "supratype" era of uncertainty was now being left behind. Henceforth long chromosomal segments apparently inherited *en bloc* were putative ancestral haplotypes or their polymorphic frozen block components thereof *until proven otherwise*. The next stages seemed clear, application of the new knowledge to improve the outcomes in transplantation and deeper molecular studies to determine the mechanism of ancestral haplotype generation and how they are conserved over many generations if not tens of thousands of years (Box 4).

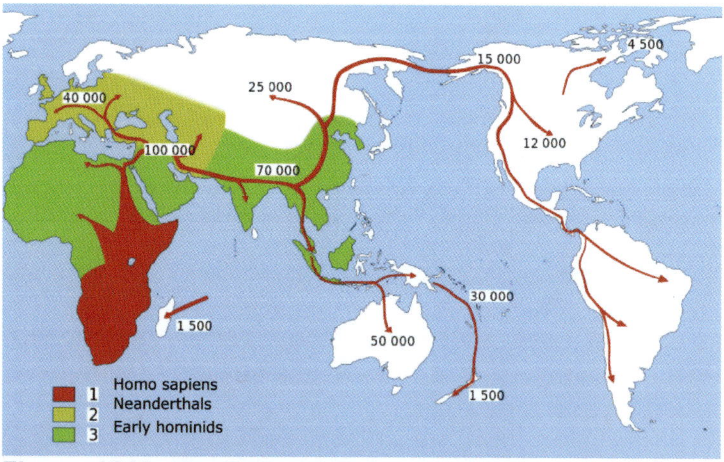

Figure 15 Human Expansion Out of Africa
http://en.wikipedia.org/wiki/Early_human_migrations. For a more accurate and detailed human migration history see the website involving the work of Professor Stephen Oppenheimer at http://www.bradshawfoundation.com/stephenoppenheimer/index.

Box 4 Note on the Age of Ancestral Haplotypes

In a study published in 1997 by Silvana Gaudieri and colleagues, employed the CL1 primers from the HLA-B TNF region in the beta block (Fig 14, Table 8) to examine certain ancestral haplotypes in Caucasoid and Mongoloid groups. Data on the migration of *Homo sapiens* from Africa generates the broad human expansion pattern shown in Figure 15. Thus at about 100,000 years ago humans began migration from Africa/Middle East into Europe and Asia.

Gaudieri et al showed that the same DNA sequences are amplified by the CLI primers from the DNA of both the 7.1AH and 7.2AH. The 7.1AH is Caucasoid specific and 7.2AH is Japanese specific. Similarly, both the 46.1AH and the 46.2AH are Asian specific (Mongoloid) yet they share the same DNA sequence between the CL primers. In both cases then the beta block HGE sequences are the same in each set of haplotypes implying first, for 7.1AH/7.2AH, that the conserved sequence predates the expansion into Japan and second, for 46.1AH/46.2AH, that the conserved sequence predates the Mongoloid split into Chinese and Japanese. Since some PFBs (the beta block) are common in more than one haplotype it implies common ancestry and suggests that subsequent shuffling and exchange with other PFBs to generate the 7.1AH and 7.2AH, as well as 46.1AH and 46.2AH. Obviously the 46.1AH/46.2AH split is recent. From the rough dates in the human expansion (Figure 15) and high frequencies of many ancestral haplotypes (e.g. Table 13) this could put the age of many (7.1, 8.1, 44.1, 57.1) dating to the beginning of the human expansion ~about 100,000 years ago. Since the 57.1AH is carried by all major racial groups - Caucasoid, Mongoloid and Negroid - it clearly predates the out of Africa migration (>100,000 years ago).

Gaudieri, S., Leelayuwat, C., Tay, G.K., Townend, D.C. and Dawkins, R.L. (1997) The major histocompatibility complex (MHC) contains conserved polymorphic genomic sequences that are shuffled by recombination to form ethnic-specific haplotypes. J Mol Evol 45 : 17-23.

Further Reading Chapter 3

Abraham, L.J., Leelayuwat, C., Grimsley, G., Degli-Esposti, M.A., Mann, A., Zhang, W.J., Christiansen, F.T. and Dawkins, R.L. (1992). Sequence differences between HLA-B and TNF distinguish different MHC ancestral haplotypes. Tissue Antigens 39: 117-121.

Alper, C.A., Awdeh, Z.L. and Yunis, E.J. (1986) Complotypes, extended haplotypes, male segregation distortion and disease markers. Hum Immunol 15: 366 – 373.

Arnand R (1986) Pulsed field gel electrophoresis: a technique for fractionating large DNA molecules. Trends Genetics 23:278-283.

Awdeh Z.L., Raum, D., Yunis, E.J. and Alper, C.A. (1983) Extended HLA/complement allele haplotypes: Evidence for T/t-like complex in man. Proc Natl. Acad, Sci. 80: 259-263.

Degli-Esposti, M.A., Andreas, A., Christiansen, F.T., Schalke, B., Albert, E. and Dawkins, R.L. (1992) An approach to the localization of the susceptibility genes for generalized myasthenia gravis by mapping recombinant ancestral haplotypes Immunogenetics 35: 355-364

Degli-Esposti, M.A., Leaver, A.L., Christiansen, F.T., Witt, C.S., Abraham, J.J. and Dawkins, R.L. (1992) Ancestral haplotypes: conserved population MHC haplotypes. Human Immunology 34:242-252.

Dupont, B. (1989) 10th International Histocompatibility Workshop -1987: Immunobiology of HLA vol 1 New York Springer-Verlag.

Egea, G.E., Yunis, I., Spies, T., Strominger, J., Awdeh, Z.L., Alper, C.A. and Yunis, E.J. (1991) Association of polymorphisms in the HLA-B region with extended haplotypes. Immunogenetics 33: 4-11.

Gaudieri, S., Leelayuwat, C., Tay, G.K., Townend, D.C. and Dawkins, R.L. (1997) The major histocompatibility complex (MHC) contains conserved polymorphic genomic sequences that are shuffled by recombination to form ethnic-specific haplotypes. J.Mol. Evol. 45: 17- 23.

McLure, C.A., Hinchliffe, P., Lester, S., Williamson, J.F., Millman, J.A., Keating., P.J., Stewart, B, J. and Dawkins, R.L. (2013) Genomic evolution and polymorphism: Segmental duplications and haplotypes at 108 regions on 21 chromosomes. Genomics 102 : 15–26.

Ohno, S. (1970) Evolution by gene duplication. George Allen & Unwin, London.

Smith, W.P., Vu, Q., Li, S.S., Hansen, J.A., Zhao, L.P. and Geraghty, D.E. (2006) Toward understanding MHC disease associations: Partial resequencing of 46 distinct HLA haplotypes. Genomics 87: 561-571.

Tokunaga, K., Sauracker, G., Kay, P.H., Christiansen, F.T., Anand, R. and Dawkins, R.L. (1988) Extensive deletions and insertions in different MHC supratypes detected by pulsed field gel electrophoresis J. Exp. Med. 168: 933-940.

Traherne, J.A., Horton, R., Roberts, A.N., Miretti, M.M. et al. (2006) Genetic analysis of completely sequenced disease-associated MHC haplotypes identifies shuffling of segments in recent human history. PLoS Genet 2 (1):e9

Wu, X., Zhang, W.J., Witt, C.S., Abraham, L.J., Christiansen, F.T. and Dawkins. R.L. (1992) Haplospecific polymorphisms between HLA B and Tumor Necrosis Factor Human Immunol 33: 89-97.

Yunis, E.J., Larsen, C.E., Ferndez-Vina, M., Awdeh, Z.L., Romero, T., Hansen,J.A. and Alper, C.A. (2003) Inheritable variable sizes of DNA stretches in the human MHC : conserved extended haplotypes and their fragments or blocks. Tissue Antigens 62: 1-20.

Zhang, W.J., Degli-Esposti, M.A., Cobain, T.J., Cameron, P.U., Christiansen, F.T. and Dawkins, R.L. (1990) Differences in gene copy number carried by different MHC ancestral Haplotypes. Quantitation after physical separation of haplotypes by pulsed field gel electrophoresis. J. Exp. Med. 171: 2101-2114.

****

Chapter 4

Grafts Matched By Ancestral Haplotypes Improve Survival

IRF/CYO research in the decade 1992-2002

Genomic Matching Technique (GMT). Genomic Matching Technique and Transplantation. "Quantal Evolution" - Mechanism of Segmental Duplication. Retroelements and Segmental Duplication. The Retroviral Induction Model of Segmental Duplication and Ancestral Haplotype Generation. Retroviral Induction Model. Genesis of Ancestral Haplotype and Complex Haplospecific HGEs? Further reading

Genomic Matching Technique (GMT)

The Perth group were now proceeding confidently into the future with a new tool for rapid genomic analysis and discovery. The ability to use a simple PCR assay to amplify short polymorphic DNA regions (<1Kb) to identify haplospecific frozen blocks (≥300Kb) or entire full length ancestral haplotypes (≥ 4 Mb) had to be exploited. It is easy, rapid and cheap. Its greatest potential would be in transplantation, particularly bone marrow transplantation (BT) otherwise termed haematopoietic stem cell transplantation (HSCT) to treat patients with various haematological malignancies. The best documented survival is where donor and recipient are full sib family members where the donor has a complete HLA Class I and Class II serological match with its sibling recipient.

However it occurred to Roger Dawkins and colleagues that a novel outcome would be to block match unrelated donor and recipients for MHC ancestral haplotypes.

GMT and Transplantation

The standard HLA serological matching in bone marrow transplantation is for a complete match for Class I and Class II MHC alleles ie. HLA-A, HLA-B, HLA-C (Class I) and HLA-DR, HLA0DQ (Class II). These MHC regions encode the peptide binding (antigen)-presentation structures important in specific T lymphocyte activation and recruitment (CD4+ve helpers/regulator T cells and CD8+ve cytotoxic T killer cells) in initiating graft versus host disease (GVHD) and thus ultimate graft rejection.

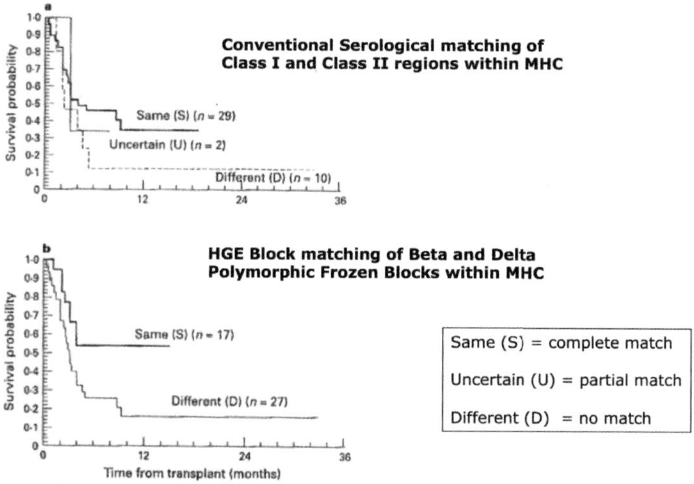

Figure 16 Matching Beta and Delta polymorphic frozen blocks of the MHC by the HGE PCR method in donor-recipient pairs significantly improves survival in bone narrow transplantation over conventional serological matching . Adapted from Figure 2 in Tay et al 1995a. Bone Marrow Transplantation 15 : 381-385.

The major obstacle in all these conventional matching procedures is not so much getting a standard serological Class I + Class II match but all the 'hidden' (unknown) transplantation antigens encoded in the non-HLA DNA segments within the MHC or even encoded at minor loci outside of the MHC e.g. regulators of complement activation on Chromosome 1 (*RCAa* locus). Such antigens can also trigger an aggressive cytotoxic T lymphocyte response leading to graft rejection.

So the Perth group set out to prove their concept by block matching entire long segments of MHC DNA to ensure the compatibility for many different protein-coding genes and non-coding regulatory RNA expressing regions (ncRNAs). It rapidly turned out that their predictions were correct and Roger Dawkins and colleagues at IRF/CYO had, in the Australian vernacular, "backed another winner"!

Soon afterwards they Patented their new donor-recipient matching method. Although with the wisdom of hindsight, over almost two decades removed from the first successful demonstration of the GMT, there has not been much monetary gain. However, the advance in basic knowledge on how to improve transplantation success has been immense and not widely recognised. The biggest surprise to the present author, and which I will deal with again, is why the GMT method has not been deployed routinely and globally to assist the survival of cancer patients undergoing unrelated bone marrow transplantation?

What has become clear, and this will be also be described shortly, is that the IRF/CYO group did influence, in part, the future course of thinking on matching suitable donor-recipient pairs. Block DNA sequencing across critical MHC regions such as the HLA-A, HLA-B/C and HLA-DP regions has now become fashionable in some clinical immunology transplantation laboratories - although not adequately prosecuted because it is likely to be too cumbersome for the rapid screening and matching of prospective donor-recipients pairs.

The basic data in the first reported use of the GMT in a retrospective trial was reported by Guan Tay and colleagues in 1995 and one aspect of the data is shown in Fig 16. Again EBV-transformed cell lines, from 45 donor-recipient pairs, were used as sources of HLA genotype/haplotypic DNA (from the Australian Bone Marrow Donor Registry Tissue Repository (27 pairs) and

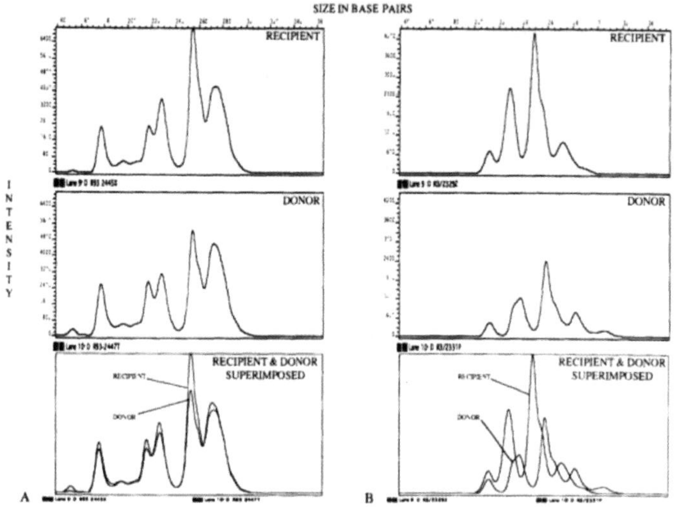

Figure 17 Matching HGE-Generated PCR Product Profiles of Delta Polymorphic Frozen Blocks of the MHC. Adapted from Figure 1 in Christiansen et al 1993. Human Immunology 38 : 42-51. Delta-Block matching showing donor-recipient patterns that are either the same (A) or different (B). The y-axis relates to band intensity and the x-axis to fragment sizes (base pairs) (A) The profiles of the donor *(middle panel)* and the recipient *(top panel)* show six major peaks at approximately 177, 196, 217, 228, 255, and 275 bp. When superimposed *(bottom panel),* the profiles of both recipient and donor are very similar (B) The profile of the recipient has peaks at 213, 230, 250. and 273 bp and the donor has peaks at 214, 239, 260, 283, and 306 bp. When superimposed *(bottom panel),* the profiles are clearly different.

the Hopital Saint Louis, Paris (18 pairs). This " was a random sample of unrelated donor pairs selected solely on availability".

As well as all other patient data the Class I and Class II HLA-DR, HLA-DQ serology was performed in the laboratory associated with the transplant unit. The

Class II HLA-DRB1 typing was done in Perth by PCR single sequence oligonucleotide typing (SSO).

In Figure 17 is shown an example of a match and a mismatch using the densitometry tracing profiles of the PCR product profiles for the delta-block of the MHC (see Fig 1). Notice that a complete delta-block match using the HGE PCR method is clear to all objective observers and a mismatch is also clear cut. Thus, it is apparent in Figure 16, that survival of unrelated bone marrow grafts matched for both the beta- and delta-blocks of the MHC can approach that expected for HLA identical sibling donors - a predicted 6 month survival of just above 50% of patients. In contrast conventional HLA Class I/II serological matching gives a 6 month survival for unrelated pairs of just above 30% of patients. In other data (not shown here) 20% of patients who are perfectly matched by HLA serology but differ for one or both of the ancestral polymorphic blocks show a 6 month graft survival.

A key point can be made here on the logistics of the donor-recipient selection step. Amongst the Caucasoid population we already know that a large percentage (≥73%) of the haplotype diversity in the MHC can be accounted for by ≥20 known high frequency ancestral haplotypes or their simple recombinants (Table 12). This means that in Caucascoids a successful matching search amongst unrelated donors for the Beta- and Delta- polymorphic blocks is a practical goal. In the presently described study by Tay et al, of the 43 pairs assessed 40% were matched at both blocks.

The use of the genomic block matching technique to vastly improve the survival of bone marrow transplants has not caught on in clinical transplantation circles - why not is the obvious question?

Apparently it has now been superseded by a superior technology with inferior results. Thus block DNA sequencing of the same MHC regions as assayed in serological Class I + Class II matching is now in fashion to search for matches. Yet, all this has revealed in preliminary work, is the enormous single nucleotide polymorphism (SNP) diversity that can exist between prospective Class I/II matched pairs (e.g. 3025 SNPs and indels ! in one study Proll et al 2011). This apparently "hidden" SNP/indel diversity is well known to the Perth group from comparison of closely related ancestral haplotypes. Thus while the concept that matching whole DNA sequences in important blocks of the MHC is now incorporated in all transplantation strategies and represents an important step forward, it is basically impractical as a screening test for the critical prospective matching of donor-recipient pairs. For this a rapid, simple and cheap test is required. The GMT using HGE-generated primer pairs fits the criteria perfectly and the present author really cannot understand why it has not been globally adopted.

Indeed block matching should be instituted for both MHC beta- and delta-blocks as well as other polymorphic gene-rich regions likely to be in epistatic interaction with the MHC, for example, those at the regulators of complement activation at the *RCAa* locus on Chromosome 1. Further, the genome-wide HGE "duplotyping" prospecting survey recently published by Craig McLure and colleagues may assist in targeting non-MHC blocks involved in transplantation rejection and graft versus host disease. However correct phasing of long MHC and non-MHC segments (i.e. the markers lie on the same chromosome) is required in those other approaches to matching that use SNP and Linkage Disequilibrium analysis (SNP/LD) and even regional DNA sequencing. These issues are avoided or minimised by HGE-block matching methods which automatically phase as they completely detect, for example with a high probability in Caucasoid populations, pre-existing ancestral haplotypes where the phase can usually be deduced from the diploid MHC genotype. In any event GMT profiles between two unrelated individuals heterozygous for their ancestral haplotypes across the MHC region can be readily distinguished by GMT profiling.

Quantal Evolution - Mechanism of Segmental Duplication

At the close of the 1980s and by the mid 1990s the following solid facts about the human MHC were established and clear to the Perth group:

First, that the MHC was characterised by extensive polymorphisms involving small and large duplications, insertion and deletion events (indels) and these structural features almost certainly underpinned the genetic "Disease Associations" characteristic of a given polymorphic frozen block within an ancestral haplotype. Much of this evidence was provided by the pulsed field gel electrophoresis technique which allowed the isolation of large chromosomal regions spanning most of the haplotypic region in question ($\geq$ 2 Mb). Thus the polymorphism was packaged within 'Polymorphic Frozen Blocks' (PFBs) of maybe $\geq$300Kb. These blocks came to be designated the alpha, beta, gamma, delta, epsilon (α, β, γ, δ, ε) blocks spanning the telomeric Class I (HLA-A, HLA-B), through the central non-HLA complement region to the centromeric Class II region (HLA-DR, HLA-DQ, HLA-DP), and see Figure 1.

Second, it became clear that these PFBs were packaged as quanta within Ancestral Haplotypes of ~ 4 Mb spanning the entire MHC region. That the genome may be built on quantal packets of information was a hitherto unexpected discovery and is still not fully appreciated by the scientific community let alone the popular expositors of science in the media.

Third, since the majority ($\geq$90%) of known Caucasoid haplotypes can be accounted by maybe 100 common ancestral haplotypes or simple recombinants thereof this meant that for at least one significant human racial group the extent of known MHC diversity was documented and that this should assist the selection of donor-recipient pairs for improved bone marrow and other tissue transplantation. Other racial groups should also develop their own registry of ancestral haplotypes as quickly as possible. The extensive work by the Boston group of Yunis, Alpers and colleagues has made deep inroads here particularly for distinct racial groups living in North America.

Through the next 5 to 10 years the group focused on how gene duplication, particularly segmental duplication, could be generated to explain this extensive polymorphism both in the human and wider mammalian genome. Susumo Ohno's 1970 gene duplication paradigm from the 1970s was fully embraced and taken to its most extreme implication - the serial duplication of whole chromosomal segments (usually ≥10 - 20Kb or more).

Attention quickly turned to retroviruses and their associated retroelements as the culprits and drivers of both genomic size and DNA sequence diversification. These elements were discovered several decades earlier. At that time their general pattern and distribution in the genome was already becoming clear. The Alu repetitive retroelements, the most abundant (~ 1 million copies per genome) can be used as evolutionary time markers (Glossary Box 8). Perhaps 45% of the human genome is composed of former RNA sequences reverse transcribed into DNA and inserted back into the genome. Alu elements depend for their mobility on the reverse transcriptase activity encoded in L1 LINE elements. Glossary boxes 5 through 9 provide basic information about these elements.

A known feature about the smaller retroelements, Alu and L1, is they often become transcriptionally active and, therefore, potentially mobile as retrotransposons when cells are under stress: in vitro tissue culture, retroviral and other viral infections, heat, various chemicals and possibly by general inflammatory cytokine signals from activated bystander cells in the immune and haematopoietic systems.

Indeed Jerzy Kulski and colleagues speculated around this very point in 2000. Thus " the clusters of retroelement activity that we see in the genome (and the HLA class I region) most probably is partly the consequence of their actions to protect the genome's integrity in times of mutation stress and disrepair, such as during duplications or other genomic rearrangements. Thus, retroelements, like HERVs, L1 and Alu, may have a role in reverse-transcriptase-mediated repair of chromosomal breaks … thus the distribution of many transposable elements within the MHC may act also as 'bandaids' or 'scars' brought about in response to genomic stress or damage." (Kulski JK pers. comm 2014).

Retroelements and segmental duplication

Roger Dawkins and colleagues argued the following in 1999:

> " The potential importance of bursts of duplication has been recognised but the mechanism has not been resolved, and the possible relevance of indels such as human endogenous retroviruses (HERVs) should be considered. We will argue that the MHC region was captured by retroelements which generated diversity, duplication and differential disease susceptibility… and that those retroelements which invaded the MHC are included within this process of duplication and diversification before they too became frozen as components of stable ancestral haplotypes." (Dawkins et al 1999).

A decade and a half later John Trowsdale and Julian Knight reviewing the MHC make the now unsurprising comment suggesting the IRF/CYO insight has now entered the scientific mainstream:

> " Several areas of the MHC, such as the class I region, appear to have undergone repeated duplication and deletion, possibly owing to retroviral activity." (Trowsdale and Knight 2013).

Here we will describe the results of some of the main papers of the Perth group during this period which underpins this view of MHC evolution - a view not restricted to the MHC but likely to hold across the wider human and mammalian genome.

In earlier studies by others prior to the IRF/CYO engagement in the problem it was clear that repetitive retroelements could be implicated in the gene duplications noted on the multigene clusters encoding the haptoglobin and globin genes. Thus retroelement sequences were located at the borders/junctions of the duplications implying a causal role in unequal crossing over (Fig 12). The Perth group then embarked on a systematic survey of the various duplicated regions within the MHC and their associated retroelements.

In one study Jerzy Kulski and his colleagues examined a 340 Kb genomic sequence spanning the alpha-block (around Class I HLA-A region). In this new era, DNA sequencing of the MHC and many other parts of the genome was well under way as a prelude to the final assembly of the first draft of the human genome (HUGO). For detailed analyses the group used the new bioinformatics alignment tools for detecting repeat sequences, mainly retroelements, Alu (SINEs), LINEs, human endogenous retroviruses (HERVs) and long extended segmental duplications. The most useful alignment tools were the Dot Plots to detect long and short tandem duplications and RepeatMasker for detection of repetitive retroelements and DNA transposons (Dawkins et al 1999). For this region in the alpha-block, they could define the unit of duplication as not involving an individual gene but one containing several genes or gene fragments often accompanied by retroelements and various transposons (≥ 10Kb to 20Kb in size). Most importantly were the sites of the breakpoints defining segmental duplicated regions: these were found mainly *within and between* the HLA Class I/Class I - like sequences (PERB11/MIC, HLAcl1) as well as HERV-16 and other retroelements. This is suggestive evidence that they have acted as homologous template targets promoting unequal crossing over and thus tandem duplication and indel events as discussed (Fig 12).

Thus in the 62.1 ancestral haplotype there are ten genomic segments containing HLA Class I genes (J ψ, 80F, A, Kf, 16f, H ψ, G, 90f, 75f, F) arrayed, more or less in tandem in the alpha-block, as a set of serial duplications. The length and content of each copy varies because of other indel events.

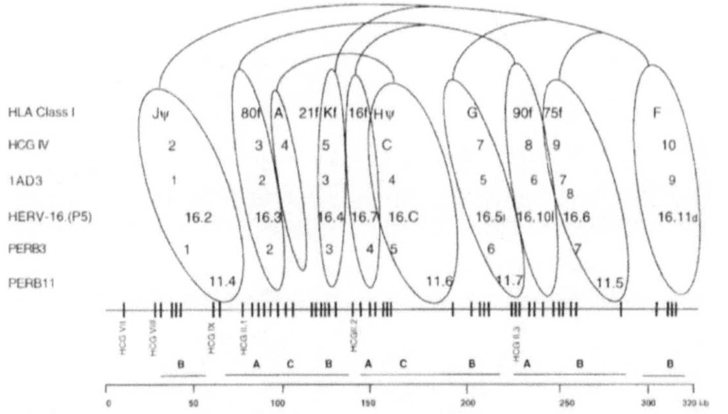

Fig 18 Imperfect segmental duplication of PERB 11, HLA, HCGIV, AD-3, HERV-16, PERB3 families in the 320 Kb region of the alpha block of the MHC . Adapted from Figure 8 in Dawkins et al 1999 Immunol Rev 167: 275-304. Full-length beads consist of PERB11, HLA, HCGIV, IAD3, HERV-16,(PS). PERB3 genes, HLA-80, HLA-A, HIA-K, HLA-16, HLA-90, and HLA-F beads lack PERB11 gene, f = fragment, 1 =LTR only, d = discontinuous. ψ =pseudogene.

Such an array "creates the impression of a string of beads......beads that vary somewhat in size, age and wear whilst sharing the same basic structure.. as the order of sequences within each unit is fixed" (Dawkins et al 1999). This picture is summarised in Figure 18 - it is a pattern that repeats itself, in principle, across all major blocks of the MHC. Littered within all the blocks are numerous retroelement repeats of all sizes, Alu (SINE), LINEs, HERVs, solo LTRs as well as DNA transposons (Boxes 5-9 in Glossary), all of which themselves, in places, have been replicated as part of longer duplications.

In another study using Alu evolutionary marker analysis (Box 8 Glossary) the group led by Kulski and colleagues (1997) established that the HLA-B and HLA-C Class I genes are the products of an extended duplication that occurred between 44 and 81 million years ago in the primate lineage well before the human-chimp separation (~6 mya). The subsequent diversification of HLA-B from HLA-C in both genomic segments appeared then to be the result of the mobility and mutation of Alu retroelements. All of these types of dated duplications and other data are entirely consistent with the Trans-Species Hypothesis of MHC evolution originally proposed by Jan Klein (the great scientist and scholar who has systemised much of our knowledge on the evolution of MHC complexes in many vertebrate species).

An overwhelming impression in reading all the papers from this period (which I call the "Gaudieri / Kulski et al papers") is their extraordinary attention to detail - and the evolutionary ages and types of retroelement and DNA transposon - layer on layer of elements often clustered and overlapping (some impacting within larger retroelements) have been laid down over evolutionary time.

In a study led by Silvana Gaudieri and her colleagues the central region of the MHC encompassing the beta- and gamma-blocks (HLA Class I, PERB11/MIC and C4 multicopy gene families) was investigated. In this region they demonstrated that segments of approximately 30 Kb had been duplicated and the main diversified feature between the duplicated arms of segments was the content of large indels of retroelements (Alu, LINE) and pseudogenes. More importantly the retroelements flanked the duplicated segments and many post-duplication insertions.

At a general level retrogenes of all types have left their impact signatures on the genome. Almost half the higher mammalian genome has been cycled through an RNA intermediate at some stage in evolution. But this is what we can easily detect - what about retrotransposition events that do not leave an obvious 'integration signature'? These I would label as 'clean integration events' mediated through targeted homologous recombination (gene conversion) and often would go completely undetected - a form of "gene replacement".

Gene conversion retrotransposition events are likely to be ubiquitous particularly amongst retroelement and endogenous retroviruses and their pathogenic, externally-derived, cousins. We indeed know already from the work of the noted retrovirologist pioneer John Coffin that a significantly large proportion, possibly a majority, of human retrovirus integration events over evolutionary time scales, are thought to occur this way. That itself suggests a general homologous recombination mechanism for retroelements. Even if one had no idea what their role might be, the sheer abundance and distribution of retroelements in gene-rich transcriptionally active regions implies they must play important roles in genome evolution.

The way the genome can increase in size via retroelement impacts is nicely illustrated by the work of Jerzy Kulski and colleagues who reconstructed the likely events over millions of years that led to a particular combination of both LINE and HERV integration events all piling on top of each other at the same site within the MHC (Figure 19). Indeed just a cursory glance at the titles of the main Gaudieri/Kulski et al papers from this period (1997-2000) conveys the story of the discovery of the high density of RE/HERV sequences within the MHC.

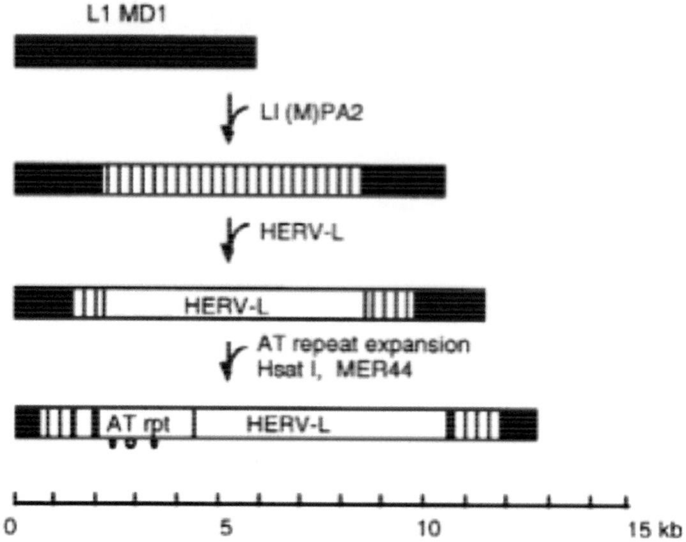

Figure 19 : Reconstruction of Events for the Integration of HERV-L. Adapted from Figure 4 , Kulski JK *et al* (1999). J Mol Evol 48: 675-683. "The scheme proposes that the region of HERV-L (PERB11) was initially spanned by a single mammalian-specific L1 (L1MD1) repeat (hatched block) about 100 mya, which was a receptor for the insertion of a primate-specific L1 (L1PA2) repeat (block with vertical bars), which in turn was the receptor for the insertion of HERV-L (open block labelled HERV-L) after the emergence of primates 65 to 85 mya. The insertion of HERV-L (PERB11) occurred between nucleotide positions 3818 and 3917 of L1(M)PA2. The HERV-L (PERB11) element was then a target site for the AT repeat expansion and insertion of Hsat1 and MER44 (denoted as circles in the AT rpt block). The insertion events have led to eventual fragmentation and deletions within the mammalian- and primate-specific L1 elements but with an overall expansion of about 8 kb within a region that was originally a L1 repeat element of about 5 kb."

A particularly good and striking example was published over several papers by Kulski, Dawkins and colleagues in 1999. Thus they analysed 656 Kb of genomic sequence obtained from alpha- and beta- blocks of the class I region using various bioinformatics sequence alignment tools particularly BLAST, Repeat Masker homology searches, dot plots, and open reading frame (ORF) analysis.

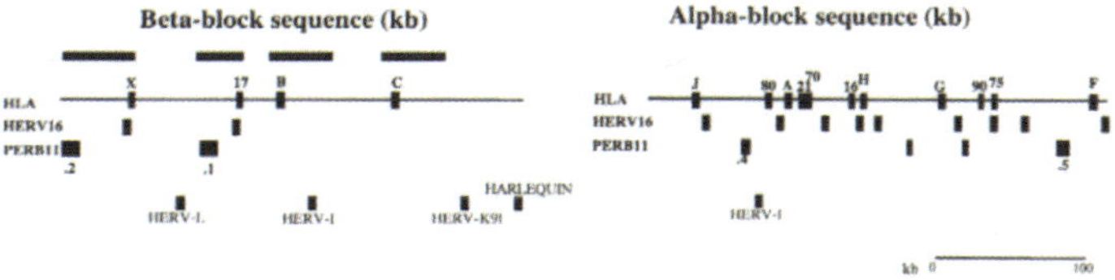

Figure 20 Map of Location of HERVs in Relation HLA-Class I and PERB11/MICA alleles. Adapted from Figure 1 Kulski JK *et al* (1999). J Mol Evol 48: 675-683.

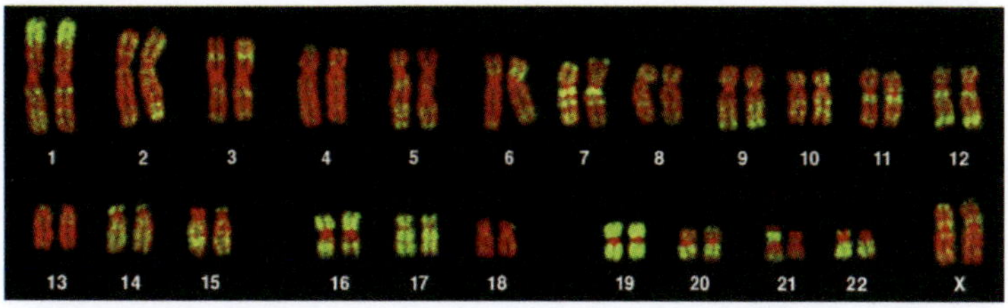

Figure 21 Distribution of ALU across Human Genome-FISH analysis. From Figure 7A in Bolzer , A. (2005) PLoS Biology 3 (5): e157. This a meta-phase spread showing the karyotype from a female human lymphocyte (46,XX). Chromosomes were hybridized with a consensus probe for Alu sequences (green) and counterstained with TORO-3 (red). Alu sequences were used as a marker for chromosomes and chromosome bands rich in genes. In general the retro-elements such as Alu (LINE) are associated in particular with, **a**. High density of protein coding genes (autosomes); **b**. High density HIV integration sites; **c**. A High density of segmental duplications. Alu sequences tend to target intronic sequences, the 5'-UTR being preferred and regulatory effects on translation efficiency (protein synthesis efficiency).

The results were striking. Thus 16 HERV sequences were detected representing at least five families : HERV-16, 11 copies; HERV-L, 1 copy; HERV-I, 2 copies; HERV-K91, 1 copy and HARLEQUIN, 1 copy. The analysis showed that the 11 copies of HERV-16 arose as a product of *serial duplication* of genomic segments containing HLA class I (HLA cI) and PERB11/MIC genes (Fig 18, 20). The other five HERVs arose after duplication probably as a consequence of single insertion events or translocations. Most of the HERV sequences had been fragmented over evolutionary time as analysis for ORFs revealed absence of intact coding sequences of the putative *gag*, *pol* and *env* gene regions typical of retroviruses whether endogenous or exogenous (Fig 20).

It is now clear from the work of the Perth group and many others that retroelement repeats are a feature of all gene-dense regions throughout the human genome. This is beautifully illustrated by the fluorescence *in situ* hybridisation (FISH) genomic pattern using a consensus Alu probe shown in Fig 21. Indeed Alu rich FISH bands are now routinely used in metaphase chromosome karyotyping as indicators of gene-rich regions! The Alu FISH pattern in Figure 21 is therefore pregnant with implications (some listed) and fully confirms the strategic decision taken by the Perth group over 20 years ago to understand the role of retroelements in MHC genome evolution. We will return to these important data later in the book.

The detailed work of John Coffin and colleagues is now pertinent in this regard, and I quote a particularly important section of their conclusions on the evolution of HERV sequences within the human and higher primate genome:

> "One surprising result is the high frequency of conversion we observed. Indeed, only two of the six loci analyzed had suffered no such events in any lineage. Solo LTRs, which arise by recombinational deletion of the intervening viral genes, and which are found by the thousands in the genomes of many animal species, are further evidence for high frequency of recombination involving ERV sequence.....The mechanism that gives rise to such events is unlikely to be provirus-specific, but probably reflects the likelihood of conversion among any repeated, nuclear DNA sequences. Because many ERVs belong to multi copy families, it is also possible that inter locus recombination gives rise to concerted evolution among some of these loci. This latter mechanism

> may explain the rather confusing topology of the RTVL-Hb tree ." (Johnson and Coffin (1999).

There is a striking yet poorly publicised fact about HERV sequences detected in human and animal genomes. Thus in a 2011 paper the Coffin group estimates that solo LTRs (mentioned above) exceed their proviral-like integrated relatives by a factor of 10 to 1! That is, more than 90% of all HERV-related DNA sequences detected by bioinformatic alignment tools in the genome are solo LTRs, most of which are likely to have arisen by "by recombinational deletion of the intervening viral genes" (LTRs are long terminal repeats bracketing the 5' and 3' ends of all infective retroviruses including HIV - they posses regulatory promoter elements for transcription (5' end initiation) and sites that also assist reverse transcription and provirus formation).

This all leads to an important, and yet surprisingly conservative, Darwinian conclusion. In the gene-rich genomic regions - in both introns and in protein-coding exons, the latter often in crippled pseudogenes - there has been built layer on layer of *numerous* retrotransposition events over evolutionary time. The major evolutionary time clock events suggest bursts of mobile RE impacts which could drive disruptive disease-inducing indels and serial segmental duplications during meiosis (Fig 18). This means that the great majority of RE impacts on the genome would have been lethal and many affected protein-coding genes would never have survived an RE burst of mobility and integration. They would have been eliminated by "purifying" selection (for programmed cell death, "apoptosis" is the default option for a cell overwhelmed by DNA sequence errors and indels). However the rare new chromosomes that did survive such RE impact events would then run the gauntlet of the rigors of meiosis, embryonic and neonatal life through to reproductive age to pass on the new chromosome with its new haplotypes to future generations. We will now deal with this important evolutionary scenario in some detail.

****

Segmental Duplications and the Birth of Ancestral Haplotypes

The late 1990s, particularly 1999, were very good years for the Perth group. They published important findings and analytical reviews. They were on the verge of understanding, at the molecular level, *how*, and most likely, *when* most ancestral haplotypes were generated during human evolution.

It is clear they had comprehensively shown that the MHC seems to have been, in their words, "captured by retroelements"! Given all that was known at that time their new model of ancestral haplotype generation, seemed an *eminently logical conclusion.* It was a very plausible hypothesis based on their data and the wider scientific literature at the time. They published what I will call their ***Retroviral Induction Model*** for ancestral haplotype generation in a major invited analytical review by Roger Dawkins and colleagues in 1999. In this author's assessment this scientific model has withstood a critical test - the test of time. It has run the potential gauntlet of lethal peer-reviewed criticism and potential refutation. None has emerged. The data on ancestral haplotypes and the RV/RE induction model has strengthened over time - indicative in the author's opinion of a *very fruitful line of scientific inquiry*. This is perhaps not surprising given that Roger Dawkins and co-workers have published widely and extensively on the topic in highly reputable peer reviewed journals for almost 35 years. The data is rock solid. It is as solid as anything can be at the molecular level in modern biomedical science. As Roger Dawkins often claims in jest in interview - it has been " one long experiment in evolution."

Thus there are no doubts on the reality of ancestral haplotypes as the basis for the quantal structure of the genome. All current extant data are consistent with such a model of genomic evolution within the MHC; and current duplotyping HGE analysis by the group suggests these rules will apply across the human genome in all gene-dense segmentally duplicated regions (McLure et al 2013).

Table 13 Summary: Facts and Assumptions on MHC Polymorphism

- The MHC is a ~ 4 Mb genomic region with blocks in the 100 kilobase range characterised by polymorphism, duplication, insertion/deletion and disease associations.

- The sheer number of diseases affected by MHC genes implies there must be multiple polymorphic functions encoded within MHC.

- The extensive genomic polymorphism is also partly redundant (duplication, indels) as the variation is often not expressed at protein level, or if it is, not important functionally.

- Segments rather than individual genes are duplicated, often imperfectly – the basic duplication unit is about 20 to 50 Kb - serial duplications of these can be 10 or more increasing copy numbers rapidly (-> large blocks ≥ 300 Kb).

- Extensive duplication can be regarded as older with expansion of segment length due to insertion.

- Insertion is wide spread and continuing. Alu, LINES and HERVs may have been inserted in bursts rather than regularly. Deletion has been frequent - resulting in major changes in primate evolution.

- Polymorphism, duplication and insertion/deletion appear related and genetic diversity quantal packets of "polymorphic frozen blocks" which are linked to form an ancestral haplotype ~ 4 Mb

As a reminder I post a summary of critical facts and assumptions Dawkins and co-workers posit as their forerunner to their retroviral induction model (Tables 13). In the overall assessment of the data there is clearly a complex relation between the polymorphic regions within MHC haplotypes and disease susceptibility which occurs in two different ways. First, susceptibilities associated with low copy number genes where there is limited polymorphism but infrequent mutation (for example 21-hydroxylase deficiency, haemochromatosis). Second, susceptibilities associated with diseases which are complex These are the most common and the outcome is a function of interactions between different polymorphic gene products which occur in higher copy number - here the effects are quantitative rather than qualitative. Thus there may be several mechanisms linking polymorphism, duplication, indels and disease. Of course a major one is the integration of HERVs and other retroelements .

The most striking feature is the apparent "quantative regulation" of the degree of disease susceptibility and the interaction of AHs. Some of this is summarised in Table 14 where I reproduce the summary of key disease associations with ancestral haplotypes published in the Dawkins et al 1999 review. Note the interactions in the diploid (heterozygous state) reveals the complexities of the genetic interactions. Clearly there has been a survival advantage as these MHC ancestral haplotypes have survived for many thousands of years (data in the table mainly refers to Caucasoids).

A model showing the evolutionary bottlenecks caused by pathogenic retroviral epidemics when a trans-species infection occurs (e.g. akin to the recent HIV / AIDS pandemic) is shown Fig 22. And in Figure 23 are summarised known and speculative events that could take place in the cell and its nucleus.

Ancestral haplotype	Locus										Disease								
	HLA-A	*HLA-Cw*	*HLA-B*	*PERB11.1*	*Bf*	*C4A*	*C4B*	*DRB1*	*DQB1*	Codon 57	IDDM	IgAS	MG	SLE	CD	PV	HH	NL	21-OH
8.1	1	7	8	01	S	0	1	0301	0201	A	++	++	++	++	++				
18.2	19	5	18		F1	3	0	0301	0201	A	++	0			0				
58.1(M)	19	?	58		S	3	0	0301	0201	A	++			+					
62.1	2	3	62	03	S	3	3	0401	0302	A	++	0	0	0					
44.2	29	4	44		F	3	1	07	0201	A	0	+	0	0	++				
57.1	1	6	57	06	S	6	1	07	0303	A	0	+	0	0	0	++			
65.1	?	8	65		S	2	1+2	0102	0501	V		++		0					+
35.2	3	4	35		F	3+2	0	0101	0501	V			0				+		
54.1(M)	?	1	54		S	2	5	0405	0401	D	++								
7.1	3	7	7	01	S	2	1	1501	0602	D	–			+	+		++		
52.1(M)	24	?	52		S	3+2	0	1502	0601	D	–								
44.1	2	5	44		S	3	0	0401	0301	D		0	0						
13.1	30	6	13		S	3	1	0700	0201										

The embellishment to the model I would add a decade and a half after its conception by Dawkins and colleagues would be this - it is now well known that retroviral infection sets up a *very mutagenic* environment within the cell and its nucleus (Fig 23). As well as triggering mobility of the retrolements themselves (LINE, Alu), the APOBEC family of cytosine-to-uracil deaminases (C-to-U) particularly APOBEC3G are activated (both as a cellular anti-viral defence and paradoxically, carried by the retrovirus particles themselves as a diversification strategy!), as well as general elevations of ADAR-deaminase mediated adenosine-to-inosine, A-to-I, mRNA editing (a cellular anti-viral defence).

Strikingly this plethora of mutagenic activities resembles many molecular aspects of the regulated process of the somatic hypermutation of antibody genes as well as their dysregulation in the genesis and progression of many somatic cell cancers.

All of this mutagenic activity would generate indels in diploid or haploid germ cell DNA (Alu, LINE and pathogenic RV), as well as mutagenic uracils (C-to-T if unrepaired), which mature on base-excision repair to mutagenic abasic sites and then if unrepaired as targets for apurinic /apyrimidimic endonucleases (APE) induced

---

Table 14 Ancestral Haplotypes and Disease Associations. Adapted from Table 4 in Dawkins et al 1999 Immunol Rev 167: 275-304. Ancestral haplotypes carry alleles which predispose to - and protect against - disease. Some ancestral haplotypes (e.g. 8.1} predispose to a number of diseases, while some diseases {e.g. IDDM) are associated with a number of ancestral haplotypes. The 7.1 haplotype appears to carry several different susceptibility genes but is also protective when in combination (heterozygous) with other AHs. Ancestral haplotype associations may be strongly (++) or weakly (+) predisposing: neutral (0); or protective (-}. Blank spaces indicate haplotypes for which disease associations have not been determined. IDDM, insulin-dependent diabetes meliitus; IgAD, IgA deficiency; MG, myasthenia gravis; SLE, systemic lupus erythematosus; CD, coeliac disease; PV, psoriasis vulgaris; HH, hereditary haemochromatosis; NL, narcolepsy; 21-OH, 21-hydroxlase deficiency.

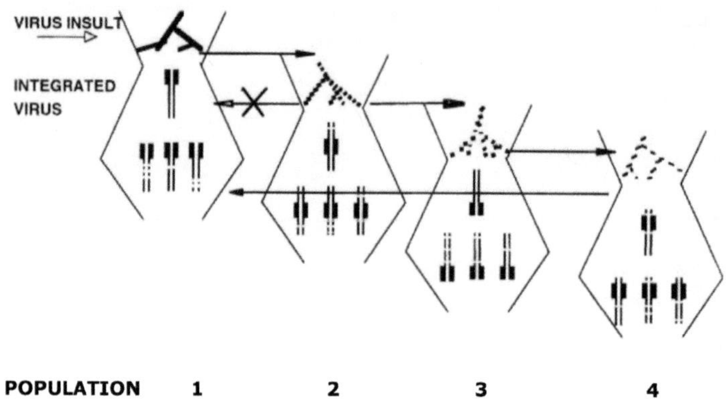

Figure 22 Retroviral Induction Model. Adapted from Figure 21 in Dawkins et al 1999 Immunol Rev 167: 275-304. A virus insult on population 1 causes a bottleneck. The invading virus mutates, represented by the expanding tree. Resistance is conferred when the virus integrates. When integrated into the MHC region, genomic diversification occurs. The integrated virus is transferred vertically with resulting expansion of the population. An arrow indicates a cross-species transfer of this virus. The same pattern is repeated in species 2. The arrow from species 2 to species 1 indicates a block in a back-transmission of the virus. The integrated virus in species 1 can still confer resistance to the virus from species 2, However, when the virus has diverged considerably from the integrated virus, that population may not be resistant. This is shown by the arrow extending from species 4 to species 1.

nicks. In addition C-to-T deaminations of methylated cytosines at CpG sites. We will return to this mutagenic RV-induced environment shortly.

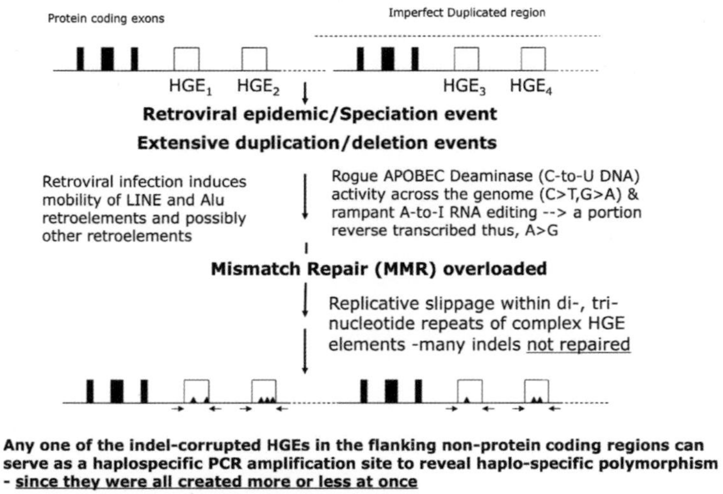

Figure 23 Retroviral Induction of Ancestral Haplotypes and Haplospecific Geometric Elements. Likely mutagenic processes within infected cells. HGE, haplospecific geometric element. The most prominent mutagenic processes are cytosine deamination of DNA and adenosine deamination of RNA. APOBEC, apolipoprotein B mRNA editing enzyme, catalytic polypeptide-like, a family of at least nine evolutionary related and conserved cytosine deaminases active on single stranded DNA (transcribed regions): APOBEC3G being the most prominent, and active in HIV infected cells as a cellular defense against retroviral infection.

****

Retroviral Induction Model

Thus in terms of the model, a trans-species pathogenic retrovirus infects a susceptible population, which then contracts creating a bottleneck. According to Dawkins and co-workers viral sequences (targeted or random) invade the MHC in germ cells of some subjects, are reverse transcribed and become integrated as proviruses and in survivors are vertically transmitted to offspring. It is predicted that as a consequence of these retroviral (and by tacit assumption, retroelement) insertions, a burst of segmental chromosomal duplication is triggered during the meiotic cycle resulting in diversification of the surviving population (see Fig 12). Thus polymorphism is generated by several mechanisms, including imperfect segmental duplication via the misaligned RV and RE elements flanking many of the MHC genes (Fig 12). Ancestral haplotypes appear, "differing from others by virtue of copy number as well as sequence polymorphism. Some of these will be relatively advantageous."

The group then made an important qualification to generalise the model. I quote in full :

> " One prediction of the model would be that HERVs are found within duplication units of other polymorphic regions within the genome. Another is that the HERVs within these polymorphic regions will bear some resemblance to retrovirus es likely to have infected ancestral populations. We already have some evidence to support these predictions. For example as shown in Fig 22, phylogenetic analysis of the MHC - encoded HERVs shows relationships to HIV1, HIV2, foamy virus and others." (Dawkins et al 1999).

This generalisation is *prescient.* It foreshadows the discovery by Craig McLure and colleagues of polymorphic frozen blocks and ancestral haplotypes in the

RCAa region of Chromosome 1 (2005) and the genome-wide duplotyping data which uncovered segmental duplications and stable haplotypes (segregation in three generation pedigrees) at 108 regions on 21 human chromosomes.

So the model was well supported at the time and it is now clear, with hindsight, and the further passage of time, that additional evidence has accrued to support it.

Thus a few years after the model was published Schroder and colleagues reported a genome-wide map of HIV integration sites in the genome of a CD4+ve T lymphocyte cell line infected *in vitro*. The results were striking. HIV favours integration in gene-rich transcriptionally active regions. These regions are also known to be rich in segmental duplications, a fact which made the genome wide HGE-driven duplotyping survey possible (McLure et al 2013). However these are also the genomic regions where (as already discussed) Alu integrations are most dense! (Fig 21). Thus the associations predicted or assumed by the model are supported by much extant evidence. Short of observing an actual speciation event taking place these are compelling data consistent with the molecular requirements of the RV/RE-induction model.

However, since then, there is now better and independent confirmatory evidence from other investigators. One study in 2001 by Hughes and Coffin reported genome rearrangements mediated by human endogenous retroviruses. In another genome-wide study Bailey and coworkers examined the flanking regions of segmental duplications and the extent of clusters of Alu retroelements near the breakpoints. Thus the following observations were reported in 2003, which I summarise in detail below:

- The human genome, in contrast to the genomes of other organisms,is enriched for large (≥10Kb) highly homologous (≥ 90% sequence identity) segmental duplications.
- The sequence features at the junctions of 9464 duplications from a genome wide set of 2366

duplication alignments were examined. This revealed a highly significant ($P< 0.0001$) enrichment of Alu short interspersed element (SINE) sequences near or within the junctions.

• Indeed just under a third (27%) of all segmental duplications terminated within an Alu repeat.

• The Alu junction enrichment was most pronounced for interspersed segmental duplications separated by ≥1 Mb of intervening sequence.

• Most strikingly - considering the work on homologous recombination amongst HERVs and their fragments (work of John Coffin just discussed above) - "Alu elements at the junctions showed higher levels of divergence, consistent with Alu-Alu–mediated recombination events."

• Classification of the Alu elements into major subfamilies was most revealing. Thus in contrast to the older Alu elements that showed no junctional enrichment, it was the younger elements (AluY and AluS) which accounted for all the enrichment.

In the light of these genome-wide data Bailey and colleagues drew the logical conclusion that:

> " ...the primate-specific burst of Alu retroposition activity (which occurred 35–40 million years ago) sensitized the ancestral human genome for Alu-to-Alu–mediated recombination events, which, in turn, initiated the expansion of gene-rich segmental duplications and their subsequent role in nonallelic homologous recombination." (Bailey et al 2003).

While this study was comprehensive (as was a later 2006 review by the same authors) I was surprised that no reference was made to the prior work or predictions of the IRF/CYO group.

All of these independently secured data are completely consistent with the retroviral induction model proposed by Dawkins and colleagues who would also propose that full length ancestral haplotypes and/or their associated polymorphic frozen blocks would also be laid down at the time of the duplicative burst or shortly thereafter. Indeed by 2006 the CYO group led by PhD student Rebecca Laird figured that since retroelements are found in all vertebrate genomes and their genomic distributions are species unique they may make ideal PCR targets from which to generate species-specific assays. The Zona Pellucida receptor 3 gene (ZP3) was targeted initially as it is responsible for sperm binding during fertilization of the oocyte (thus it retains sufficient conservation between species to enable its function but enough diversity to prevent cross-species fertilization). Then it turned out that the PCR primers were complementary to an ancient mammalian interspersed repeat (MIR) retro-element - and since it was then inadvertently shown that one primer alone could be used it allowed development of a bi-directional amplification of complementary duplicons (BACD) species-specific assay. So the involvement of retro-elements in speciation events had come full circle back to the fusion of a sperm with its egg!

Indeed the combined extant data describe, I would argue, retroelement-driven segmental duplication that is necessary to drive ancestral haplotype formation and genomic evolution.

But what is our level of scientific confidence that the extant observational data describe events that may have occurred more than 100,000 years ago in the small interbreeding groups of our remote ancestors who left Africa to give rise to all current human beings?

Let me use a quantitative scientific analogy from astrophysics. Although we have never witnessed the genesis of a segmental duplication/ancestral haplotype generation event I hold we can be as confident of their reality by a mechanism like that proposed, as we are confident that the Meteor Crater in Arizona was caused by the impact of a large meteorite about 50 metres across striking about 50,000 years ago (Fig 24).

Meteor Crater is a meteorite impact crater approximately 37 miles (60 km) east of Flagstaff, and 18 miles (29 km) west of Winslow in the northern Arizona desert of the United States. The crater was created about 50,000 years ago. The object that excavated the crater was a nickel-iron meteorite about 50 meters (55 yards) across. The speed of the impact was at 12.8 kilometers per second (28,600 mph). It is believed that about half of the impactor's bulk was vaporized during its descent before it hit the ground. Impact energy has been estimated at about 10 megatons. The meteorite was mostly vaporized upon impact, leaving little in the crater

Figure 24 Meteor Crater.
http://en.wikipedia.org/wiki/Meteor_Crater

****

Genesis of Ancestral Haplotypes and HGEs

Ancestral haplotypes present a *profound* challenge to our thinking about evolution as they have been maintained with *little or no change* throughout the human expansion and possibly for at least 50,000-100,000 years. Some like the 57.1AH are far older, in circulation in Africa prior to the expansion. Although some changes occur within microsatellites and other replicative slippage hotspots it is still amazing that little changes have occurred elsewhere and there has been conservation of genomic polymorphism.

While the mechanisms are still not understood, recombination is suppressed within PFBs yet *recombination* does occur at the borders which *shuffles polymorphic frozen blocks* between ancestral haplotypes rather than changes within a block. In the early part of this book I discussed the evidence that recombination hotspot motifs probably border PFBs. The motif PRDM9 appears to control activation of mammalian recombination hotspots. Sequence and length differences may also lower the probability of a cross over. I think it more likely that families of motifs regulate recombination i.e. the process is not haphazard or random as implied in in the theory of population genetics - certainly not in vertebrates and the mammalian line leading to human beings.

With respect to suppression of point mutations and short indels the most likely DNA repair candidate mechanism, in my opinion, is sister chromatid strand-directed mismatch repair (Lester, S, Dawkins, R.L. pers.comm). One likely overarching process could involve the integrated control of DNA repair enzymes within a block-wide large Transcription Factory. Peter Cook and colleagues have contributed much to functional genome structure in their characterisation of Transcription Factories. And one can imagine that in actively transcribed gene-rich regions stoichiometric concentrations of DNA repair proteins and supramolecular assembles would be very high to deal promptly with all DNA (and RNA) errors including polynucleotide strand recombinations occurring within the factory. Genes not under the control of a frozen block (for what ever reason) would be expected to incur a higher incidence of unrepaired spontaneous mutation and free recombination.

How are the different ancestral haplotypes or extended polymorphic frozen blocks generated? My argument has been that whatever event(s) created them they were created *in their entirety at once*, with all their characteristic duplications, indels, gene copy number variations (cnv) and SNPs. We can say this is because of the discovery *and pragmatic success* of HGEs in block matching in transplantation and in prospecting for polymorphic stable blocks across numerous segmentally duplicated regions in the genome. That is, a given PFB if not an entire ancestral haplotype can be *completely* defined by a short (200-400bp) complex DNA sequence - haplospecific geometric element (HGE) - present, at least in two copies, in the two arms of a large tandem imperfect segmental duplication. Thus two different and widely spaced sets of HGEs within the beta-block of the MHC detect the same ancestral haplotypes (Fig 25). These were separately designed and used as haplospecific markers almost a decade apart by different members of the IRF/CYO group. This to me points to something special about how long mutated stretches of DNA are laid down during ancestral haplotype formation. A plausible scenario based on mismatch repair overload is outlined in Fig 23 and above.

Accordingly any HGE (in principle) in a given region e.g. in a HLA-B duplicated region, is suffice to mark the entire region. This means that multiple different HGEs in that stretch are laid down *at once* by, for example, replicative slippage and deletion at the complex repeat/geometric element and impaired mismatch repair (MMR) due to all the other mutagenic DNA sequence errors (C-to-U, abasic sites, ssDNA nicks etc) requiring

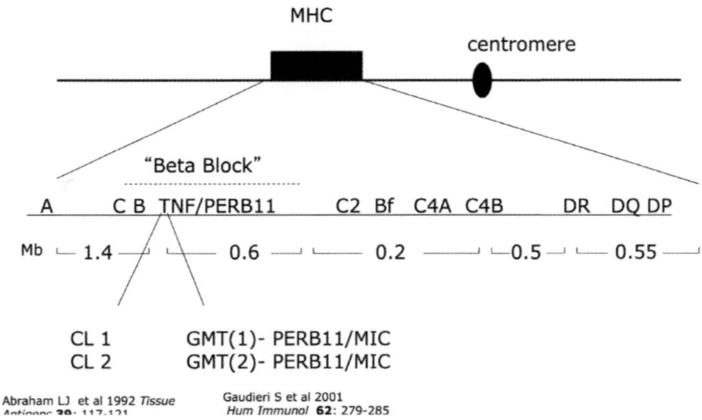

Figure 25 The number of different AHs can be defined entirely by two different HGE/PCR primer pairs in different regions of the beta block. Indeed blocks of similar ancestral ages may have been laid down at once such as the gamma and delta blocks in Chinese and Japanese populations studied by Gaudieri and colleagues (see Box 4).

attention during the meiotic crisis associated with a "speciation event". Therefore it is assumed that the overload in MMR occurs because the amount of damaged DNA requiring repair before replication far exceeds the stoichiometric levels of MMR complexes within the affected germ cell. An overloaded, and thus impaired, MMR mechanism, would be expected during retroviral crises, a stress-related process which would also produce population bottlenecks during speciation events.

So we have a well supported model and plausible molecular explanations. Where next for the IRF/CYO group?

Further Reading Chapter 4

Bailey, J.A., Liu, G. and Eichler, E.E. (2003) An Alu transposition model for the origin and expansion of human segmental duplications Am. J. Hum. Genet. 73: 823-834.

Bailey, J.A. and Eichler, E.E. (2006) Primate segmental duplications: crucibles of evolution, diversity and disease. Nat. Rev. Genetics 7: 552-564.

Baudat, F., Buard, J., Grey, C., Fledel-Alon, A., Ober, C., Przeworski, M., et al. (2010) PRDM9 is a major determinant of meiotic recombination hotspots in humans and mice. Science 327: 836 -840.

Bolzer, A., Kreth, G., Solovei, I., Koehler, D., Saracoglu, K. et al. (2005) Three-dimensional maps of all chromosomes in human male fibroblast nuclei and prometaphase rosettes. PLoS Biol 3(5): e157.

Burns, M.B., Temiz, N.A and Harris, R.S. (2013) Evidence for APOBEC3G mutagenesis in multiple human cancers. Nat Genet. 45: 1-7.

Christiansen, F.T., Tay, G., Smith, L.K., Witt, C.S., Petersdorf, E.W., Bradley, B. and Dawkins, R.L. (1993) Histocompatibility matching for bone marrow transplantation donor-recipient pairs in the 4AOHW cell panel Human Immunology 38 : 42-51.

Cook, P.R. (2010) A model for all genomes: the role of Transcription Factories. 395: 1-10.

Cordaux, R. and Batzer, M.A. (2009) The impact of retrotransposons on human genome evolution Nat Rev Genet 10: 691-703

Cheung, V.G., Sherman, S.L. and Feingold E. (2010) Genetic control of hotspots. Science 327: 791-792.

Dawkins, R.L., Leelayuwat, C., Gaudieri, S., Tay, G., Hui, J., Cattley, S., Martinez, P. and Kulski J. (1999) Genomics of the major histocompatibility complex: haplotypes, duplication, retroviruses and disease Immunological Reviews 167: 275-304.

Erickson, L.M., Kim, H.S. and Maeda, N. (1992) Junctions between genes in the haptoglobin gene cluster of primates. Genomics 14 : 948-958.

Fitch, D.H.A., Bailey, W.J., Tagle, D.A. ,Goodman, M., Sieu, L. and Slightom, J.L. (1991) Duplication of the gamma-globin gene mediated by L1 long interspersed repetitive elements in an early ancestor of simian primates, Proc Natl Acad Science USA, 88 : 7396-7400.

Gaudieri, S., Dawkins, R.L., Habara, K., Kulski, J.K. and Gojobori T. (2000). SNP profile within the human major histocompatibility complex reveals an extreme and interrupted level of nucleotide diversity. Genome Research 10: 1579-1586.

Gaudieri, S., Giles, K.M., Kulski, J.K. and Dawkins, R.L. (1997) Duplication and polymorphism in MHC: Alu generated diversity and polymorphism within the PERB11 gene family. Hereditas 127: 37-46.

Gaudierei, S., Leelayuwat, C., Tay, G.K., Townend, D.C. and Dawkins, R.L. (1997) The major histocompatibility complex (MHC) contains conserved polymorphic genomic sequences that are shuffled by recombination to form ethnic-specific haplotypes. J. Mol. Evol. 45: 17- 23.

Gaudieri, S., Leelayuwat, C., Townend, D. C., Kulski, J. K. and Dawkins, R.L. (1997) Genomic characterization of the region between HLA-B and TNF: Implications for the evolution of multicopy gene families J Mol Evol 44: S147-S154.

Gaudieri, S., Kulski, J.K., Balmer, L., Giles, K.M., Inoko, H. and Dawkins, R.L.(1997) Retroelements and segmental duplications in the generation of diversity within the MHC. DNA Sequence 8: 137-141

Gibbs, A.J. and McIntyre, G.A. (1970). The Diagram, a method for comparing sequences. Its use with amino acid and nucleotide sequences. Eur. J. Biochem 16: 1-11.

Goldfarb, T. and Lichten, M. (2010) Frequent and efficient use of the sister chromatid for DNA Double-Strand Break Repair during budding yeast meiosis. PLoS Biol 8(10): e1000520. doi:10.1371/journal.pbio.1000520

Hughes, J.F. and Coffin, J.M. (2001) Evidence for genome rearrangements mediated by human endogenous retroviruses during primate evolution. Nat Genet. 29: 487-489.

Johnson, W.E. and Coffin, J.M. (1999) Constructing primate phylogenies from ancient retrovirus sequences. Proc. Natl. Acad. Sci. (USA) 96: 10254-10260.

Jones, R.B., Song, H., Xu, Y., Garrison, K.E., et al (2013) LINE-1 retransposable element DNA accumulates in HIV-1-infected cells. J. Virol. 87 : 133307-13320.

Kitcharoen, K., Witt, C.S., Romphruk, A.V., Christiansen, F.T. and Leelayuwat, C. (2006) MICA, MICB, and MHC beta block matching in bone marrow transplantation: relevance to transplantation outcome. Human Immunology 67 : 238-246.

Klein, J. (1987). Origin of major histocompatibility complex polymorphism: the trans-species hypothesis. Human Immunology 19: 155-162.

Klein, J http://en.wikipedia.org/wiki/Jan_Klein

Klein, J., Sato, A., Nagl, S. and O'hUigín, C. (1998) Molecular trans-species polymorphism Ann Rev. Ecol. System. 29 : 1-21.

Kulski, J.K, Gaudieri, S., Martin, A. and Dawkins R.L. (1999) Coevolution of PERB11 (MIC) and HLA class I genes with HERV-16 and retroelements by extended genomic duplication J. Mol Evol 49: 84–97.

Kulski, J.K., Gaudieri, S., Inoko, H. and Dawkins, R.L. (1999) Comparison between two human endogenous retovirus (HERV)-rich regions within the major histocompatibility complex. J. Mol Evol 48: 675-683.

Kulski, J.K. and Dawkins, R.L. (1999) The P5 multicopy gene family in the MHC is related in sequence to human endogenous retroviruses HERV-L and HERV-16. Immunogenetics 49: 404-412.

Kulski, J.K., Gaudieri, S., Bellgard, M., Balmer, L., Giles, K., Inoko, H. and Dawkins, R.L. (1997) The evolution of MHC diversity by segmental duplication and transposition of retroelements J. Mol Evol 45: 599-609.

Kulski, J.K., Gaudieri, S. and Dawkins RL. (2000) Transposable elements and the metamerismatic evolution of the HLA class I region. In Major Histocompatibility Complex. Evolution, Structure and Function, ed., M. Kasahara, Springer-Verlag, Tokyo pp.158 -177.

Laird, R., Gaudieri, S., Berry, J., Williamson, J., Yang, J-S. and Dawkins, R. (2006) Vertebrate species profiling in one step using a single primer DOI: 10.1385/Forensic Sci. Med. Pathol. 2 :3:165

Lester, S., McLure, C., Williamson, J., Bardy, P., Rischmueller, M. and Dawkins, R.L. (2008) Epistasis between the MHC and the RCAα block in primary Sjogren syndrome. Ann. Rheum. Dis. 67: 849-854.

Lindley, R.A. and Steele, E.J. (2013) Critical analysis of strand-biased somatic mutation signatures in TP53 versus Ig genes, in genome wide data and the etiology of cancer ISRN Genomics. Vol 2013 Article ID 921418, 18 pages (Open Access online)

Longman-Jacobsen, N., Williamson, J.F., Dawkins, J.F. and Gaudieri. S. (2003). In polymorphic genomic regions indels cluster with nucleotide polymorphism: Quantum Genomics. Gene 312: 257-261.

McLure, C.A., Hinchliffe, P., Lester,S., Williamson, J.F., Millman, J.A, Keating, P.J., Stewart, B.J. and Dawkins, R.L. (2013). Genomic evolution and polymorphism: Segmental duplications and haplotypes at 108 regions on 21 chromosomes. Genomics 102: 15-26.

McLure, C.A., Williamson, J.F., Smyth, L.A., Agrawal, S., Lester, S., Millman, J.A., Keating, P.J., Stewart, B.J. and Dawkins. R.L. (2005) Extensive genomic and functional polymorphism of the complement control proteins. Immunogenetics 57: 805-815.

Myers, S., Bowden, R., Tumian, A., Bontrop, R.E., Freeman, C., MacFie, T.S., et al. (2010) Drive against hotspot motifs in primates implicates the PRDM9 gene in meiotic recombination. Science 327: 876-877.

Parvanov, E., Petkov, P.M. and Paigen, K. (2010) Prdm9 controls activation of mammalian recombination hotspots. Science 327: 835

Proll, J., Danzer, M., Stabentheine, S., Niklas, N., Hack, C., et al (2011) Sequence capture and next generation resequencing of the MHC region highlights potential transplantation determinants in HLA identical haematopoietic stem cell transplantation DNA Res. 18: 201–210.

Ryan, F. (2009) Virolution. Collins, London.

Schroder, A.R.W., Shinn, P., Chen, H., Berry, C., Ecker, J.R. and Bushman, F. (2002) HIV-1 integration in the human genome favors active genes and local hotspots Cell 110: 521-529.

Schulz, J., Leese, F. and Held, C. (2008) Introduction to Dot Plots
/http://www.code10.info/index.php?option=com_content&view=article&i...n-to-dot-plots&catid=52:cat_coding_algorithms_dot-plots&Itemid=76

Sciamanna, I., Alberto Gualtieri. A., Cossetti. C., Osimo, E.F. et al (2013) A tumor-promoting mechanism mediated by retrotransposon- encoded reverse transcriptase is active in human transformed cell lines Oncotarget, www.impactjournals.com/oncotarget/Published October 14, 2013 (Open Access)

Steele, E.J. (2009) Mechanism of somatic hypermutation: Critical analysis of strand biased mutation signatures at A:T and G:C base pairs. Molec. Immunol 46 : 305-320.

Steele, E.J., Williamson, J.F., Lester, S., Stewart, B.J., Millman, J.A., Carnegie, P., Lindley, R.A., Pain, G.N.,and Dawkins, R.L. (2011) Genesis of ancestral haplotypes: RNA modifications and RT-mediated polymorphism. Human Immunology. 72 : 283-293

Subramanian, R.P., Julia H Wildschutte, J.H., Russo, C. and Coffin, J.M. (2011) Identification, characterization, and comparative genomic distribution of the HERV-K (HML-2) group of human endogenous retroviruses

Retrovirology 2011, 8:90
http://www.retrovirology.com/content/8/1/90

Tay, G.K., Witt, C.S., Christiansen, F.T., Charon, D., Baker, D., Herrmann, R., Smith, L.M., Diepeveen, D., Mallal, S., McCluskey, J., Lester, S., Loiseau, P., Teisserenc, H., Chapman, J.,Tait, B. and Dawkins. R.L. (1995a). Matching for MHC haplotypes results in improved survival following unrelated bone marrow transplantation. Bone Marrow Transplantataion 15 : 381-385.

Tay, G.K., Witt, C.S., Christiansen, F.T., Corbett, J.M. and Dawkins, R.L. (1995b). The identification of MHC identical siblings without HLA typing, Experimental Hematology 23 : 1655-1660.

Trowsdale, J. and Knight, J.C. (2013) Major Histocompatibility Complex Genomics and Human Disease Annu. Rev. Genomics Hum. Genet. 14: 301-323

Witt. C., Sayer, D., Trimboli, F., Saw, M., Herrmann, R., Cannell, P., Baker, D. and Christiansen. F. (2000). Unrelated Donors Selected Prospectively by Block-matching Have Superior Bone Marrow Transplant Outcome, Human Immunology 61: 85-91.

Yunis, E.J., Larsen, C.E., Ferndez-Vina, M., Awdeh, Z.L., Romero, T., Hansen, J.A. and Alper, C.A. (2003) Inheritable variable sizes of DNA stretches in the human MHC : conserved extended haplotypes and their fragments or blocks. Tissue Antigens 62: 1-20.

Chapter 5

Paradigm Shift : Quantal Structure Of The Genome

IRF/CYO Research in the decade 2002 - 2013

Quantal Structure of the Genome. The CYO Foundation shifts to rural Canning Vale. Extending the Ancestral Block Concept to Chromosome 1. Duplotyping Across the Human Genome. Ancestral Haplotypes in the Canine, Boney Fish and Bovine Genomes. Ancestral Block Haplotypes and Human Individuality. Human Genome - Part rediscovery of "Block Haplotype Concept" Evaluation of Genomic Data. A Digression into Extended Haplotype Homozygosity. Is There a Role for "Natural Selection"? Functional Equivalence of Ancestral Haplotypes. Natural or Artificial Selection Under an Ancestral Haplotype Paradigm. Interactions of Ancestral Haplotypes in Heterozygotes. Further reading.

Quantal Structure of the Genome

During this period the Perth group relocated from the Department of Clinical Immunology at Royal Perth Hospital and the Queen Elizabeth II Medical Centre (University of Western Australia) to the southeast outskirts of Perth, to the sandy melaleuka country of the "Melaleuka Cattle Stud" owned by Roger Dawkins. His vision to establish a privately funded scientific research campus began to be implemented in then rural Canning Vale. The new research site with an associated CYO Tavern in the context of a small research village with science / technology innovation incubators was set up by the C.Y. O'Connor ERADE Village Foundation. A small group of talented PhD students and post docs helped initiate the campus in the early 2000s. The C.Y. O'Connor ERADE Foundation is named in honour of Roger Dawkins' great-grandfather (Box 10).

Box 10 **C.Y. O'Connor**

It is relevant in this volume on the history of the discovery of ancestral haplotypes to briefly make a personal digression into Professor Roger L Dawkins' own ancestry. The original C.Y. O'Connor was Roger Dawkins' great-grandfather - a heroic figure in the establishment of the State of Western Australia at the turn of the 19th - early 20th century. He is now widely admired and lauded in WA as a founding father. C.Y. O'Connor was a man of big ideas and the resilience and strength of character to see them through to completion against formidable political opposition and physical obstacles. Accordingly, as an engineer and builder of grand projects, C. Y. O'Connor was chosen to lead the construction team that laid the 300 mile Perth to Kalgoorlie water pipe so that water could be supplied to the goldfields around Kalgoorlie and Coolgardie on the semi arid Nullabor Plains. As *The Australia Pipeliner* put it in a 2007 article " The proposal was greeted with derision. Who had ever heard of such a thing? Pump 5,000,000 gallons per day to Coolgardie, a distance of over 300 miles, through 30 inch steel pipes? The idea was absurd, impossible!" I am sure the reader will grant me the licence to draw a likely conclusion from this ancestry on his mother's side: Roger Dawkins clearly inherited the visionary and pioneering spirit from his great-grandfather, maybe as a set of maternal ancestral haplotypes.

Extending the Ancestral Block Concept to Chromosome 1

The Perth group pushed ahead using the new laboratory and seminar rooms located at the CYO Foundation. Over the past decade at CYO they have made a series of very important advances all of which extended the quantal genomic structure across both the human and higher mammalian genomes.

In a series of important papers Craig McLure and colleagues extended in detail the PFB and AH concepts and disease associations to another human chromosome likely to be in *trans epistasis* (regulatory interaction) with the alleles within the MHC on Chromosome 6. They investigated in detail the genomic structures on chromosome 1 which contains loci for the multi gene families encoding complement control proteins (CCPs) involved in the regulation of complement activation (RCA) alpha block at chromosome band 1q32 (RCAα block). Again they used Dot Plot techniques to search for regions of extended segmental duplications. Then they looked at the DNA sequence in the non-protein coding regions for complex elements - potential haplospecific geometric elements (HGEs) - so that appropriate PCR primers could be designed to prospect for polymorphism at the HGE sites and reveal ancestral haplotypes as the group showed earlier for the MHC. PCR products were compared within three-generation families and haplotypes assigned. The common haplotypes uncovered *all segregated in unrelated three-generation families.* They described more than 20 new ancestral haplotypes at the RCAα block which covers a segmental duplication of approximately 350Kb comparable to the block sizes within the MHC. They showed differences amongst the AHs in their significant associations with various diseases, recurrent spontaneous abortion (RSA) and psoriasis vulgaris (PV) as well an epistatic associations between 01AH and Sjogren syndrome (pSS) with HLA-DR3. The 15 most common ancestral haplotypes in their study population of three-generation families (644 chromosomes) accounted for approximately 70% of the population studied.

Thus the picture being painted here is a rerun of the experience already gained from studying polymorphic frozen blocks within the MHC. Indeed the authors comment again on the power of the HGE approach used so successfully as the genomic matching technique (GMT) in bone marrow transplantation:

> " The present study provides another example of the utility of the GMT approach. This simple procedure has demonstrated linked polymorphisms

> including at least one of functional significance (Birmingham et al 2003). Short of sequencing and somehow assembling hundreds of kilobases in at least 30 subjects, we know of no other approach which could reveal more than 20 different haplotypes with such extensive polymorphisms. " (McLure et al 2005).

Whilst whole genomic sequencing and regional whole exome DNA sequencing has got faster and cheaper these past 10 years it is still too cumbersome and cost ineffective compared to the GMT/HGE technique for revealing ancestral haplotype polymorphism in human populations rapidly and on a large scale in multiple DNA samples.

Thus within the RCAα block the polymorphisms (variation) in the genetic information around and within the protein coding genes in the regulation of complement activation (CR1, CR1-like, CR2, MCP, MCP-like) is packaged or concentrated as polymorphic "quanta". Indeed, across the PFBs of the MHC as a whole, information is packaged as "quanta". And when one examines polymorphisms in terms of single nucleotide polymorphisms (SNPs) this is also concentrated as "quanta" associated, surprisingly, with indels clusters. And between any two MHC ancestral haplotypes the SNP profiles are discretely packaged - extreme frequencies of interrupted levels of SNP density.

A picture is emerging that clustered polymorphism in regions of dense segmental duplication maybe the norm across the genome. This quantal information is packaged within ancestral haplotypes. The genomes in human populations appears to have a polymorphic quantal structure - of conserved polymorphism from remote ancestors.

So the first draft of the human genome reported in 2000 was truly a rough draft of one composite genome. In the current era of thousands of whole genomes being sequenced each year (usually in diseased patients e.g.

various cancers) *it is imperative that we get the conceptual picture right* on how polymorphism is packaged across the genome. Indeed, as I will discuss below, major mistakes were made in the first 5 years of the whole genome sequencing era (2000-2005) and these mistakes are still having an impact on how genomic data is interpreted today.

Somatic Mutation Cancer Paradigm

Some of this misguided and confused thinking can be traced to the somatic mutation-selection paradigm of cancer which has infected all our thinking about evolutionary change. The author has been as guilty as most in feeding this paradigm which is suitable for its context. In a well written article on the cellular and tissue architecture influencing the role of somatic mutation during the development of cancer Frank and Nowak state: "Thus high mutation rates speed adaptation in novel environments by providing greater variability, the engine of evolutionary change." However ancestral germline mutations are inherited as stable polymorphisms: they have already run the gauntlet of evolution and are important because they have been integrated into the already established gene circuits and interactions of the whole organism.

****

Duplotyping Across the Human Genome

I have already discussed various aspects of the McLure et al 2013 study because of the great importance of that work. The full impact of that study was being digested by the author as this history was being drafted and is thus the most recent report from the Perth group at the time of writing.

Recall the group used the HGE-based "duplotyping" technique to establish numerous potential AH blocks across most of the extensively duplicated regions (n=108) on 21 chromosomes. Thus the polymorphism appears roughly related to both gene-density and density of segmental duplication across the genome. It is important to also recall that polymorphic PCR products were *only* scored in the McLure et al study if the haplotype patterns could be shown to be stably transmitted (segregated) in a three generation family. All stably transmitted haplotypes are all *potential ancestral haplotypes* and a wider population study is now required. Thus the polymorphism uncovered in that study was enormous: *no haplotype pattern was repeated* at any region across 30 unrelated subjects of diverse ethnicity nor in two other unrelated three generation families. A wider study would confirm the AH status for many of the PFBs in the different duplicated regions (i.e. repeat PCR product pattern in individuals from unrelated families) and extend the number to other potentially polymorphic conserved regions.

The deep evolutionary history of the PCR primer sites was revealed when polymorphism was also found in cross species amplification - they showed primer pairs designed to amplify human duplicons can also amplify the syntenic sites in other species to variable extents in revealing polymorphism.

Ancestral Block Haplotypes and Human Individuality?

McLure and colleagues demonstrated that there are at least 108 block haplotypic regions across 21 chromosomes of the human genome. From their data they make the conservative assumption that there are 100 polymorphic ancestral haplotypes at each block and meiotic recombination occurs randomly between blocks such that:

> " .. each of the possible genotypes at each block will be associated randomly with each at other blocks. In a given individual there are 10^{400} possibilities. Clearly, there are more than sufficient possible genotypes to account for the individuality of the 8 billion humans alive currently and also of all their ancestors. Contrariwise, to account for human individuality, it is only necessary to postulate that there are tens of independently segregating ancestral haplotypes at tens of blocks. Clearly there is more polymorphism than has been appreciated previously." (McLure et al 2013).

These combinatorial numbers are truly of cosmic proportions and provide much food for thought. It leads to the possibility that the MHC, given its role in preserving the body against infectious diseases and other bodily disorders, is the *master genome controller* (R.L Dawkins pers. comm.). This is not unreasonable given that human individuality would need to apply across the genome, pointing to the MHC as a central coordinator of the physical integrity of the body and the individual self.

Ancestral Haplotypes in the Canine, Bony Fish and Bovine Genomes

The Perth group have also extended in detail the PFB and ancestral haplotype concepts to the MHC of other mammalian species including the dog MHC region and to an innate immunity-determining region encoding a polymorphic mannose binding lectin (MBL) region in Zebrafish. More recently the group has successfully taken the ancestral haplotype/PFB concept into the beef cattle industry by investigating the association of MRIP ancestral haplotypes ($\geq$14Mb) associated with healthy beef (low melting point fat) and present on bovine Chromosome 19. This ongoing study is certain to have an impact on genetic selection techniques for domesticated animals in the beef cattle, sheep and related industries.

Human Genome - Part rediscovery of "Block Haplotype Concept"

I have given a preview of parts of this problem at the start of the book. However the problem runs a little deeper and will be dealt with here and then again in the section on the phenomenon of "peer-group silence" in the face of fundamental discovery. I will be assisted here by a good contemporary review of the genomics of the major histocompatibility complex and human disease by John Trowsdale and Julian Knight

By the early 2000s, as the first draft of the human genome was put down the HapMap project began based on determining single nucleotide polymorphisms (SNPs) across different ethics groups and setting up a large database of human genetic diversity. A conceptual basis of such an approach had just been published by the group led by Mike Goddard. Then, at this time, several high profile groups in the US and Europe published in the journals *Nature* and *Science* definitive genome-wide confirmations of both the extensive linkage dysequilibria and apparent block structure across much of the human genome. In other words, without citing the prior work of the Perth or Boston groups, the genomics research community appeared to confirm the definitive work of these groups as well as the wider multicenter HLA/MHC collaborations of the previous 20-30 years, which actively nurtured the discovery of the block haplotype structure of the human MHC. It is also unclear whether the groups using these SNP/LD analyses fully understood the biological and medical significance of polymorphic ancestral haplotypes currently existing at high frequencies in human populations. It is thus now clear to many informed objective observers of this period in biomedical science that this claim of credit and priority was premature as the interpretation based on SNP/LD analysis has turned out to be wrong in its details. In contrast to what had been established within the MHC by the Perth and Boston groups the "blocks" discovered by such analyses were all small on average ~ 50Kb.

It represents a good example of the inappropriate application of the limited methods and concepts of neo-Darwinian population genetic to the far more complex evolved structure of the human and wider mammalian genome.

SNP/LD analysis is based on the fact that if two adjacent SNPs are inherited together they are said to be "linked" and in so called " linkage dysequilibrim". A picture then, in theory, can be built of linkage groups by pair-wise comparisons of SNPs. The aim, with increased density of SNPs is to do genome-wide studies where a disease or quantitative/qualitative trait tagged with a SNP may assist in locating where the relevant genes are in the genome. The approach is statistical and based on the assumptions of neo-Darwinian population genetics which I outlined at the start of this book.

The exceptional study is that of the group of Daniel Geraghty and John Hansen at the Fred Hutchinson Cancer Research Center in Seattle which clearly understands that SNPtags need to be located in the context of existing long *phased* MHC ancestral haplotypes.

Historical and analytical reviews of the priority in this area are overdue. This is one reason why I am writing the history of how ancestral haplotypes were discovered. It is necessary to state this all here because little, if any, scientific credit has been properly apportioned to the Perth or Boston groups despite the fact that, in the case of the Perth group, it had just published (a year or two beforehand) a *major invited review* of all their key findings in the prestigious journal *Immunological Reviews* (Dawkins et al 1999). Indeed the Boston group felt moved to redress this by publishing a clarification review of the history of the field in 2003 in the authoritative journal, *Tissue Antigens* (Yunis et al 2003). Then in a critical analysis they showed, as has already been made clear, that GWA studies with SNPtags do not detect known conserved extended or ancestral haplotypes within the MHC and are therefore of dubious value in pin-pointing polymorphic MHC alleles associated with genetic diseases (Alper et al 2006).

****

Evaluation of Genomic Data

It should now be clear that genomic data of any type cannot be evaluated with an uncritical mind once the scientist has *confronted* the data on ancestral haplotypes. I underline the word confronted because it is every scientist's duty to do so - it is the only reliable way for mankind in differentiating scientific knowledge from mythology, delusion and superstition.

If there is one message I hope to leave the reader with, it is this: the quantal or ancestral haplotype structure of the genome changes all the rules on how genetical data should be interpreted, and how selective breeding programs with livestock should be prosecuted. While the Gaussian bell curve is an important statistical method and assumption, the scientist must understand both its power and limitations.

I have stressed that particularly suspect are those genome-wide association studies (GWAS) where the aim is to link a disease predisposition, or some other qualitative characteristic, with a particular genetic locus tagged with a SNP. Recall that this scepticism is due to the astounding and confounding fact that the MHC, at chromosome band 6p21.3 and now many other regions of the genome appear composed of large polymorphic frozen blocks. Within such blocks there are large numbers of variant genetic "alleles" and there is no recombination or ongoing mutation (apart from microsatellite mutation hot spots) and any given ancestral haplotype is associated with ill-defined sets of polymorphic genes determining susceptibility to complex diseases. The scientist cannot use Gaussian-based population statistical methods on such data let alone view it through the ideological prism of neo-Darwinian population genetics.

The conventional expectation of SNP/LD analysis of the HapMap data, particularly within the well studied Caucasoid population for genome-wide SNP-linkage research, is that many of the evenly spaced SNPs across the

MHC should at least pick up most of the common linked alleles within a known ancestral haplotype. However, as we know, and to our surprise, this is not the case as clearly shown by the critical analysis of Alper and colleagues (the square brackets contain clarifying remarks):

> " The inherent inability of LD analysis to "see" DNA fixity beyond three [SNP] markers contributes to the failure of SNP/LD analysis to define in detail or even detect CEHs [conserved extended haplotypes or ancestral haplotypes] in the MHC and probably elsewhere in the genome. More importantly, the use of statistical analysis, rather than direct haplotype determination [by three generation family segregation analysis] and counting, fails to reveal the details of haplotype structure essential for gene localisation. Given the oversimplified picture of the MHC (and probably the rest of the genome) provided only by SNP/LD-defined blocks, it is questionable wheth er this approach will be of great help in disease susceptibility gene localisation or identification." (Alper et al 2006).

Further, the analytical algorithms associated with GWA SNP-LD analyses have been set up to process "edited" data that avoid runs of SNPs i.e. the SNPs need to be evenly spaced for LD analysis. Runs and clusters of SNPs can be common between two closely related MHC ancestral haplotypes but the actual sequence is specific for each.

A quantal distribution of packets of DNA sequence polymorphism is the norm. Thus, to reflect reality, a complete non-edited set of SNPs should be employed in GWA studies. Further, to reinforce what was put at the start of this book, the data imply that most, if not all, current SNPs occurring at reasonable frequencies *already exist in close linkage disequilibrium within a known ancestral haplotype or PFB fragment.* Finally, when recombination occurs, it takes place between frozen blocks and it is not random. This fact undermines a key assumption of LD-SNP studies which assume a more or less even and random frequency of free recombination across the length of a given chromosome (so that recombination hot spots can be detected). There seem to be deep and fatal flaws in all GWA studies involving SNPs. As Alper and colleagues conclude:

> " We believe the problem with SNP/LD analysis is the statistical nature of the approach rather than the use of SNPs. .... It is the observed haplotypes that must be the basis of any rational approach to gene localisation." (Alper et al 2006)

There is an ironic commonality between some of the genetic resources used by the SNP/LD analytical approach and the direct ancestral haplotype/CEH approach of the Perth and Boston groups. Thus, as an example, in one large scale high resolution SNP/LD study across the human MHC, Miretti and colleagues selected SNP markers which were typed for genetic stability "..across a panel of 190 DNA samples from CEPH families (Utah residents with ancestry from northern and western Europe); DNA was obtained from the Coriell Cell Repository." These same CEPH families are used in ancestral haplotype analyses by the Perth group. The approaches and conclusions, however, are utterly different. In the first case, the inheritance within a family confirms the SNP is indeed a germline polymorphism at that genomic site and in the latter, the haplotype segregation pattern confirms the Mendelian inheritance and stable segregation of a very long ancestral chromosomal segment into the megabase range.

The latter fact has been taken into account in the important SNP discovery and MHC re-sequencing study by Geraghty, Hansen and their colleagues (Smith et al 2006). Thus the 46 distinct MHC haplotypes analysed were either derived from homozygous HLA typed cell lines or based on HLA haplotypes defined by family pedigree studies; in this way they could deduce the ancestral haplotypes for almost all their SNP data and the correct phase (they lie on the same chromosome), thus associating the SNP with known HLA allele information. GWAS SNP/LD studies that do not incorporate all this phased information will be misleading and most often fail to detect an ancestral block sequence.

Can we find other genome-wide SNP data independent of - not tainted by - the data used by the HapMap consortium? Before going into these data it is important to recount important developments that are going on within some of the groups who analysed this type of genomic data.

Indeed several of the main groups associated with the earlier SNP/LD statistical analyses have now published papers which recognize these problems. Thus Leslie and colleagues have now based such analyses on the fact that very long extended ancestral haplotypes *exist within the MHC* and the contiguous runs of characteristic SNPs *within such haplotypes* need to be factored in for methods trying to predict the identity of Class I and Class II alleles in, for example, searches for donor - recipient match ups for bone marrow transplantation. This is certainly an improvement and as Trowsdale and Knight summarise " .. approximately 100 SNPs were needed to predict HLA alleles to two digits with ~ 95% accuracy." In the field of SNP/LD genome wide analyses in improvement of livestock breeding programmes for desired complex phenotypic traits Wray and other members of the Goddard group have now published cautionary tales on the "pitfalls of prediction analysis and show how naïve implementation can lead to severe bias and misinterpretation of results."

Trowsdale and Knight point out another problem with using SNPs particularly in relation to identifying disease causing loci and other associations within the MHC region. It concerns a problem which has been recognised for many years (since mid 1980s) about the genetic diversification mechanism for the MHC genes encoding Class I and Class II peptide (antigen) presenting structures. It explains the huge number of HLA Class I and Class II alleles ($\geq$1000) circulating in the global human population. The question is, when did they arise? Are they the result of ongoing germline mutation and natural selection? Given the reality of the dominance of MHC ancestral haplotypes spanning the entire $\geq$ 4 Mb region in any population group (e.g. well studied Caucascoids) ongoing germline mutation is unlikely.

It has also been clear for some time that these genes could be involved in a very special type of gene conversion targeting mechanism during cell division in both germ line cells and somatic cells. This allows the conversion and swapping of localised genetic information (~ 25 to 90 bp) encoding the peptide binding grooves in such antigen presenting structures. This process could potentially happen at each generation and lead to new HLA Class I and Class II genes. This would generate rare new HLA alleles within the context of a longer ancestral haplotypic region. It is conceivable that this mechanism is speeded by orders of magnitude up during major species crises and during speciation events as envisaged under the Dawkins et al Retroviral Induction Model - that is, it would be part of past evolutionary genetic processes which have generated ancestral polymorphic frozen blocks.

In the author's opinion a far more likely biological role for groove-site targeted gene conversion within the MHC is that it is a *somatic genetic strategy* beneficial to the host - much like the somatic hypermutation process targeted to antibody variable genes following stimulation by foreign antigens leads to potentially higher affinity antibodies to thwart disease. Thus during times of extreme cellular stress within the cell (cancer, viral infection) the MHC gene conversion mechanism is activated and allows the cell to position new, and now "foreign", Class I and Class II molecules on their surface. These would be natural and immediate targets for aggressive (alloreactive) cytotoxic T cells leading to that cell's elimination from the body in much the same way a foreign tissue graft is rejected. Put in historical perspective this would have the same result as McFarlane Burnet's famous "Immune Surveillance" process against cancer cells but with a mechanistic twist.

A Digression into Extended Haplotype Homozygosity

I will now delve into a hitherto unexpected area of genomics. I asked the question can we find other contemporary evidence based on genome-wide distributions of SNP data which *do detect* very long ancestral DNA segments which are polymorphic?

The author has searched for data and analyses that are independent in approach of the main genomic research groups who worked initially interpretating the HapMap data and advising how the database should be set up. I discovered this when researching signatures of "positive natural selection" at the molecular level of the genome and then stumbled across "extended homozygous haplotypes".

I then asked how common is extended haplotype homozygosity (EHH) across the genome? My common sense expectation was that they should be most evident in inbred populations. I also thought that they are likely to be regions we expect to be highly conserved thus necessary in all healthy cells - that is, they harbour "house keeping genes" which all cells need to stay alive running all the fundamental biochemical pathways. There would be obligate and intense intracellular positive selection to conserve them as well as their usually *cis* linked highly conserved regulatory regions (e.g. small and long non-protein coding regulatory ncRNAs).

The first hint on how relevant such studies might be came from the work of Gibson and colleagues. They studied the HapMap SNP data for unrelated parents across the various ethnic groups. They used algorithms to search for long runs of homozygosity in the genome (where the inherited SNPs from paternal and maternal chromosomes are identical). This has its drawbacks but as a bare minimum it will give some idea of "homozygosity". They looked for homozygous tracts with a density of one SNP per 5 Kb and lengths >1 Mb. That is, about 200 contiguous SNPs that are identical (homozygous).

To their surprise they found many such tracts and showed they were directly related to the recombination pattern. Maximum tract lengths ranged from 6 to 18 Mb. For example amongst 60 unrelated CEPH Utah residents with ancestry from northern and western Europe (CEU) there were 498 tracts uncovered in the genome. Their "Discussion" of the data is most illuminating. Thus "Homozygous tracts were found in regions of extensive LD and locally low rates of recombination." However in the original HapMap reports .. only the "presence of relatively short haplotype "blocks" were found by LD analysis. Clearly, they conclude, the presence of much longer homozygous tracts (≥ 1 Mb tracts) was perhaps not widely anticipated in those original HapMap reports.

Gibson and colleagues concluded that long homozygous tracts of conserved DNA are common in the genome and then stated :

> " It is conceivable that the abundance of homozygous regions and their contribution to long regions of high LD will significantly reduce our ability to fine map disease genes using association." (Gibson et al 2006)

However there was no mention anywhere in their paper of polymorphism in long tracts and blocks.

It is to another EHH study led by David Curtis and colleagues we now turn. This study addressed the issue of extend haplotype polymorphism in the genome and reports, unexpectedly, on a genome-wide picture *entirely consistent* with the discovery of ancestral or extended conserved MHC haplotypes by the Perth and Boston groups since the early 1980s.

Thus Curtis and colleagues used 1411 subjects from America and Netherlands and living subjects all unrelated individuals, and three data populations pooled. They were all genotyped for 312,316 SNPs in the context of a GWA study for late onset Alzheimer's diseases. The dataset was downloaded from the Translational Genomics Research Institute website.

And Curtis et al wrote their own software to systematically identify regions of extended homozygosity, defined as containing a minimum number of contiguous, genotyped homozygous SNPs (≥10) and as extending over a minimum physical long distance (≥1 Mb). In contrast to HapMap analyses no criterion was set for minimum SNP density.

The Curtis group found many regions of extended homozygosity (outside of centromeres) - and very common on the X chromosome - more common than previously appreciated. And their estimate of extended haplotype population frequencies ranged from their lower end detection level of 2% to the more common ones at the 80% level. They also noted at many loci extensive SNP polymorphism in different haplotypes at the same >1Mb locus - many others I suspect went undetected by their method as these would be below the detection threshold (determined by the population size).

They found that regions of extended homozygosity over 1Mb are common with an average of 3.56 regions per chromosome per subject, and containing on average 73 homozygous SNP markers per haplotypic region. They noted that they were concentrated at chromosomal regions where presumably recombination is rare and they have a markedly non-random distribution.

Thus they infer in effect that long polymorphic haplotypes with fixed ancestral recombination boundaries are common in the human genome where they have fixed locations. There must also be recombination "hot spots" at the ancestral block boundaries.

All of this is entirely consistent with the quantal structure of the genome advanced by the IRF/CYO group. As indicated I take all these results as third party confirmation from an entirely different data source. Inspection of the original data in the paper is recommended. Thus in commenting on their data they state :

> ".. (the patterns) seem to be a consequence of some haplotypes being very common in the population and although sometimes this reflects the effect of a very common haplotype there are examples of two or three common haplotypes at the same location, each very different from each other at several SNP positions."
> (Curtis et al 2008)

However what is most interesting is their speculation on the evolutionary genetic mechanism of generation. Curtis and colleagues think such patterns can only be understood if a contiguous run of variant SNPs is laid down at once:

> " ..From a theoretical point of view one might observe this if there were a mechanism whereby several or many mutation events could occur in a common haplotype simultaneously... for example if there were some very local change in chemical microenvironment around part of a chromosome leading to mutagenesis"…. Alternatively..this could happen if there were two or more separate ancestral populations in which the haplotypes had evolved independently .. But this does not coincide with current thinking about human ancestry."(Curtis et al 2008)

Because their study population is six times larger than the HapMap project and focuses largely on Caucasoids it is relevant to the IRF/CYO data. Thus in their EHH pattern for chromosome 6 on the short arm at about position 30Mb (where the MHC would be located) there is a clear set of homozygous tracts. My estimate by visual tract counting suggests ~ 50 EHH tracts. This suggests that the homozygous frequency is 3.5% (50/1411) and a global or "generic" ancestral haplotype frequency of about 19% for this region. This number is in the expected range from ancestral haplotype frequencies in the total Caucasoid population reported in the IRF/CYO study by Degli-Esposti and colleagues in 1992.

So like the naïve members of "Plato's cave" emerging into the light, the lessons learnt from the genomics of the MHC are not dancing shadows but point to a tangible quantal genome structure in apparently unrelated data secured for a quite different purpose.

Is there a role for "Natural Selection"?

Throughout this book I have identified and highlighted the major IRF/CYO break-throughs and put them in historical context. The idea that the human and higher primate (and mammalian) genome is composed of very large extended stretches of DNA which have remained stable for thousands of years has profound implications. It fundamentally alters our thinking about evolutionary mechanisms and the origins of human genetic diversity and individuality. Indeed the MHC re-sequencing study by the Geraghty and Hansen Seattle group has confirmed the *complete DNA sequence identity* of many independently isolated and phased MHC haplotypes e.g. the 8.1AH. Aly and others have confirmed the complete identity of the many SNPs across many different 8.1AHs in a separate study. Clearly, the approach to defining AHs using the far simpler and cheaper GMT/HGE technique has been comprehensively confirmed by these separate DNA sequencing and SNP identity studies – it therefore should be the clinical method of choice in MHC block matching donors and recipients in bone marrow transplantation.

Table 1 and Box 1 and the associated text at the start of this book summarises the popular and mainstream scientific view which assumes ongoing gene mutation and natural selection as the drivers of recent human evolution. But the IRF/CYO data tell another story. For many genetic regions the genesis of diverse variations (polymorphisms) in immunological and other functions occurred *many thousands of generations ago i*n the higher primate lineage of human evolution – these long variant DNA sequences have been passed down largely unchanged from remote human ancestors. This really is a daunting molecular discovery about human origins.

We all should tip our hats to the persistence and resilience of Roger Dawkins and colleagues in Perth and Chester Alper and Edmond Yunis and their colleagues in Boston for sticking to their guns.

All the data imply that the key 'natural selection' and gene testing events occurred aeons ago probably during major speciation or genetic crisis episodes - and, given our experience with the recent HIV pandemic, these crises are likely to have been caused by pathogenic retroviruses causing considerable numbers of deaths and significant population bottlenecks leaving smaller founder populations with multiple new germline integrated retroviral sequences capable of driving unequal meiotic crossover duplication and deletion (Fig.12a,b). These surviving founder populations however, as Jan Klein and others have estimated, were likely to be of reasonable size, maybe from 1,000 up to 10,000 intra-breeding individuals.

Speciation from the Chimpanzee was almost certainly prolonged and messy and several crises may have punctuated the split. No doubt the divergences – or speciation-like events - within the human lineage, from the early hominids, through Neanderthals and emergence of *Homo sapiens* was also messy. As we have recounted the IRF/CYO group put forward their RV/RE-induction theory based on their genomic data showing that the MHC is rich in both retroviral and retro-element insertions that would both cause DNA sequence polymorphisms (insertion and deletion events or indels) and promote tandem segmental duplications. The 'natural selection' events therefore occurred at the genesis of the current crop of known ancestral haplotypes many thousands of years ago. The 'survival of the fittest ' via genomic diversity ever since seems to have been achieved by recombination and shuffling of these large stable preexisting DNA blocks.

Functional Equivalence of Ancestral Haplotypes

By this reckoning all extant MHC ancestral haplotypes can be considered *functionally equivalent* in terms of survival value. Given their associations with complex diseases they all appear to be - as recognised many years ago in population genetics as 'balanced polymorphism' situations - *genetic compromises* : the beneficial nature of the sum total of the polymorphic alleles and their expression as variant proteins within a block is compatible with life.

These therefore are very surprising and unsettling observations at odds with mainstream neo-Darwinian and population genetic views of evolution. This may partly explain why they have not been emphasised even by those who have confirmed the observations in a general way by the imprecise and yet misleading SNP/LD studies. It also may explain why these profound concepts on genome structure have been tacitly ignored by genome-wide association studies using SNPtags. Such SNP/LD studies pay no heed to the likelihood that the SNPs discovered, as we have stressed, are almost certainly associated already with their own ancestral haplotype. To show this in a SNP/LD study real *extra work* is required (as demonstrated by Geraghty and Hansen and their colleagues) to show long range linked segregation of SNP markers in three generation family studies to confirm the reality of such a potential block ancestral haplotype defined by SNP-LD criteria.

But *surely,* I can hear the cry, " natural selection' in terms of identifiable environmental selection forces *must* have played a role in maintaining the structural genetic integrity of ancestral haplotypes at high frequency, even if only a reinforcing one over the millennia? In the deep evolutionary past leading to the present can we identify recurrent life threatening episodes applying strong environmental selection pressures?

Certainly the necessity and broad ability to mount recurrent immune responses to recurrent and common bacterial and viral diseases would be one selective force. But what about other conserved disease susceptibility alleles within the MHC which play no role in immunity - how are they preserved? And why?

To identify two possible examples, apparently non-immunity related, we can consider Hereditary Haemochromatosis (HH) strongly associated with the common 7.1 MHC ancestral haplotype; and Type 1 Insulin Dependent Diabetes Mellitus (IDDM) strongly associated with the common 8.1 MHC ancestral haplotype. In the case of HH and iron overload it is possible to imagine many occasions where infectious diseases and uncontrolled wound bleeding would have favoured the continued selection of an "iron-withholding defence mechanism ". In the case of high blood glucose it is possible to imagine a cryoprotective adaptation which may have protected northern Europeans in past ice ages (Moalem et al 2005; Smith et al 2006). It may also be possible to identify plausible evolutionary origins of other apparently maladapted MHC associations e.g. 21-hydroxylase deficiency and the 47.1 AH.

Natural or Artificial Selection ? - Under an Ancestral Haplotype Paradigm

It is clear that it is hard to identify simple natural selection gene targets. The current set of ancestral polymorphic haplotypes had all their alleles tested long ago. It is the shuffling of this conserved polymorphism which is important for each new generation.

Under the new quantal genetic paradigm, selection for a single gene is very likely to co-select the ancestral haplotype carrying that allelic variant! Is there any extant evidence consistent with such a phenomenon. There must be many frustrated livestock breeders who having selected for one trait have been confounded by getting a collection of traits in the new "genotype' which were unintended.

It is possible that the Lactase Persistence gene (LCT) was selected this way 10,000 years ago when dairy farming and milk consumption commenced - it appears to have been selected in the context of a very long (>1 Mb) haplotype sweeping down through the 3' end of the gene for at least 1 Mb. These LCT haplotypes are often homozygous in European populations (and others, displaying the EHH effect recounted above). LCT haplotypes have appeared to reach high frequency this way although it is also possible that the 3' sweep brought along LCT-specific regulatory regions.

But since many other genes would be selected *en bloc* by such a positive selection process then the possibility exists of the ancestral genes now causing disease in an inappropriate environment. However out-breeding could help ancestral haplotypes fragment by recombination. How would this occur mechanistically? That is, how does outcrossing promote recombination? Certainly in the studies on ethnicity and ancestral haplotypes Dawkins and colleagues often talk about this possibility. On racial admixture they comment:

> "..our data suggest that ancestral haplotypes are maintained if they exceed some threshold of frequency, but they fragment and perhaps even disappear if they fall below this threshold, as might occur after racial admixture. The high degree of conservation of ancestral haplotypes observed in relatively homogeneous populations may reflect reconstitution by preferential recombination within regions with identical genomic sequence. The maintenance of ancestral haplo-

types may reflect the survival of a distinct population group with a finite number of different ancestral haplotypes, but a restricted amount of racial admixture." (Degli-Esposti et al 1992)

One possibility is that at meiotic chromosomal alignment hot spot recombination regions, now at quite *different positions* on the ethnic distinct paternal and maternal chromosomes, misalign much like the proposal for unequal crossing over (Fig 12) leading to aberrant crossing over following chiasmata. It is also conceivable that there is a connection between appropriate and inappropriate homologue pairing and the final DNA-to-DNA recombination step steered by base pairing. Xu and Cook discuss the role of similarities between active transcription factories in the longer range pairing process prior to actual DNA-DNA base pairing crossover. In the context of interracial mating slight misalignments in homologue pairing during meiosis could lead to promotion of recombinational shuffling at PFB borders and ancestral haplotype fragmentation.

Interactions of Ancestral Haplotypes in Heterozygotes

In the same way that a SNP or allele of a single copy gene does not exist in isolation the same applies for MHC ancestral haplotyes and their interactions with *other* genes within other ancestral haplotypes and other gene products from other blocks across the wider genome. The diploid composition of all somatic cells means that there is a contribution of genetic expressions from both paternal and maternal chromosomes. How do they interact in a heterozygote? Do they clash negatively, positively or are they neutral in their phenotypic expressions?

In most heterozygotes it might be surmised in advance that the outcome will be benign. This is because of our conclusion that all MHC ancestral haplotypes are "functionally equivalent" on evolutionary grounds (ie. they have survived many thousands of generations). It often does not matter as often haplotype heterozygotes are usually in the majority in any population. However several illustrative examples of strong interactive effects on the strength of disease associations in heterozygotes can be made (Table 15).

Thus individuals carrying the 8.1 ancestral haplotype are strongly predisposed to developing juvenile onset Type 1 diabetes (IDDM). In combination with 44.2 and 57.1 this predisposition is undiminished. Yet in individuals where the 8.1 is in combination with 7.1 or 52.1 (M) the disease *never emerges* suggesting a strong inhibitory and thus protective effect of the these two ancestral haplotypes and the development of insulin dependent juvenile onset Type I diabetes. As an aside the 8.1AH is generally considered an all round "High Responder" in immunological terms and the elevated levels of autoimmune-related diseases in 8.1 may be related to their general "High Responder" phenotype.

There are also clear epistatic interactions between ancestral haplotype blocks of the RCAα block and the HLA-DR3 allele of the MHC (almost certainly associated with the 8.1 AH). Thus, autoantibody positive patients with primary Sjogren's syndrome are in very strong epistatic interaction with AH 1 in the RCAα block.

Closing comment

I have decided to end the book here. I could go on about prescriptive lists for scientific analyses in the new quantal paradigm. But I have decided to let the historical narrative of the discovery of ancestral haplotypes speak for itself.

Further reading Chapter 5

Alper, C.A., Larsen, C.E., Dubey, D.P., Awdeh, Z.L., Fici, D.A. and Yunis, E.Y. (2006) The haplotype structure of the human major histocompatibility complex. Human Immunology 67: 73-84.

Altshuler, D., Brooks, L.D., Chakravarti, A., Collins, F.S., Daly, M.J. and Donnelly, P. (Members of writing group: The International HapMap Consortium.) (2005) A haplotype map of the human genome. Nature 437: 1299 - 320.

Aly, T.A., Eller, E., Ide, A., Gowan, K., Babu, S.R., Erlich, H.A., Rewers, M.J., Eisenbarth, G.S. and Fain, P.R. (2006) Multi-SNP Analysis of MHC Region Remarkable Conservation of HLA-A1-B8-DR3 Haplotype. Diabetes 55: 1265-1269.

Bersaglieri, T., Sabeti, P.C., Patterson, N., Vanderploeg, T., Schaffner, S.F., Drake,J.A., Rhodes, M., Reich, D.E. and Hirschhorn J.N., (2004) Genetic signatures of strong recent positive selection at the Lactase gene Am. J. Hum. Genet. 74: 1111-1120

Burnet, F.M. (1970) Immunological Surveillance, Oxford: Pergamon Press.

Curtis, D., Vine, A .E. and Knight J (2008) Study of regions of extended homozygosity provides a powerful method to explore haplotype structure of human populations. Annals of Human Genetics 72: 261-278

Dawkins, R.L., Leelayuwat, C., Gaudieri,S.,Tay, G., Hui, J., Cattley, S., Martinez, P. and Kulski, J. 1999. Genomics of the major histocompatibility complex: haplotypes, duplication, retroviruses and disease Immunological Reviews 167: 275-304.

Degli-Esposti, M.A., Leaver, A.L., Christiansen, F.T., Witt, C.S. Abraham, L.J. and Dawkins, R.L. (1992). Ancestral haplotypes: conserved population MHC haplotypes. Human Immunology 34: 242-252.

Frank, S.A. and Nowak , M.A. (2004) Problems of somatic mutation and cancer. BioEssays 26: 291-299.

Gabriel, S.B., Schaffner, S.F., Nguyen, H., Moore, J.M., Roy, J., Blumenstiel, B., et al. (2002) The structure of haplotype blocks in the human genome. Science 296: 2225–2229.

Gaudieri, S., Dawkins, R.L., Habara, K., Kulski, J.K. and Gojobori.T. (2000) SNP profile within the human major histocompatibility complex reveals an extreme and interrupted level of nucleotide diversity. Genome Research 10: 1579-1586.

Gibson, J., Morton, N.E. and Collins, A. (2006) Extended tracts of homozygosity in outbred human populations Human Molecular Genetics 15 : 789-795

Hughes, A.L. and Nei, M. (1988) Pattern of nucleotide substitution at major histocompatibility complex class I loci reveals over domiant selection. Nature 335 : 167-170.

Hughes, A.L. and Yeager, Y. (1998) Natural selection and the evolutionary history of major histocompatibility loci. Frontiers in Bioscience 3, d509-516 (May 26 1998).

Jackson, A.N., McLure, C.A., Dawkins, R.L. and Keating, P.J. (2007) Mannose binding lectin (MBL) copy number polymorphism in Zebrafish (*D. rerio*) and identification of haplotypes resistant to *L. anguillarum.* Immunogenetics 59:861–872

Klein, J. (1996) The major histocompatibility complex and the neutral theory: the exception that proves the rule? J. Genet. 75: 117-124.

Leslie S, Donnelly P. and McVean G. (2008). A statistical method for predicting classical HLA alleles from SNP data. Am. J. Hum. Genet. 82: 48-56

Lester, S., McLure, C., Williamson, J., Bardy, P., Rischmueller, M. and Dawkins, R.L. (2008) Epistasis between the MHC and the RCA alpha block in primary Sjogren syndrome, Ann. Rheum. Dis. 67 : 849-854.

Longman-Jacobsen, N., Williamson, J.F., Dawkins, R.L. and Gaudieri, S. (2003) In polymorphic genomic regions indels cluster with nucleotide polymorphism: Quantum Genomics. Gene 312 : 257-261.

Martinsohn, J.T., Sousa, A.B., Guethlein, L.A. and Howard, J.C. (1999) The gene conversion hypotheis of MHC evolution : a review. Immunogenetics 50 : 168-200.

McLure , C.A., Hinchliffe, P., Lester, S., Williamson, J.F., Millman, J.A., Keating., P.J., Stewart, B, J. and Dawkins, R.L. (2013) Genomic evolution and polymorphism: Segmental duplications and haplotypes at 108 regions on 21 chromosomes. Genomics 102 :15-26

McLure, C.A., Williamson, J.F.,Smyth, L.A., Agrawal, S.,Lester, S., Millman, J.A., Keating, P.J., Stewart, B.J. and Dawkins. R.L. (2005a) Extensive genomic and functional polymorphism of the complement control proteins. Immunogenetics 57: 805-815.

McLure, C.A., Kesners, P.W., Lester, S., Male, D., Amadou, C., Dawkins, J.R., Stewart, B.J., Williamson, J.F. and Dawkins, R.L. (2005b). Haplotyping of the canine MHC without the need for DLA typing, International Journal of Immunogenetics 32 : 407-411.

McLure, C.A.,Williamson, J.F., Stewart, B.J., Keating , P.J. and Dawkins, R.L. (2004) Genomic analysis reveals a duplication of eight rather than seven short consensus repeats in primate CR1 and CR1L: evidence for an additional set shared between CR1 and CR2 Immunogenetics 56: 631-638

Meuwissen, B. Hayes, J. and Goddard M.E. (2001) Prediction of Total Genetic Value Using Genome-Wide Dense Marker Maps Genetics 157: 1819-1829

Miretti, M.M., Walsh, E.C., Ke, X., Delgado., M. et al. (2005) A high-resolution linkage-disequilibrium map of the human major histocompatibility complex and first generation of tag single-nucleotide polymorphisms. Amer. J. Hum. Genet. 76: 634-646.

Moalem, S., Storey, K.B., Percy, M.E., Peros, M.C. and Perl. D.P. (2005) The sweet thing about Type 1 diabetes: a cryoprotective evolutionary adaptation. Medical Hypotheses 65 : 8-16.

Nathenson, S.G., Geliebter, J., Pfaffenbach, G.M., and Zeff, R.A. (1986) Murine major histocompatibility complex class-I mutants: molecular analysis and structure - function implications. Ann. Rev. Immunol. 4: 471-502.

Smith, W.P., Vu, Q., Li, S.S., Hansen, J.A., Zhao, L.P., and Geraghty, D.E. (2006) Toward understanding MHC disease associations: Partial resequencing of 46 distinct HLA haplotypes. Genomics 87: 561-571.

Stumpf, M.P.H. (2002) Haplotype diversity and the block structure of linkage disequilibrium. Trends Genet 18:226 –228.

Trowsdale, J. and Knight, J.C. (2013) Major Histocompatibility Complex Genomics and Human Disease Annu. Rev. Genomics Hum. Genet. 14: 301-323

Wall, J.D. and Pritchard, J.K. (2003) Haplotype blocks and linkage disequilibrium in the human genome. Nat Rev Genet 4:587–597.

Weinberg, E.D. (1993). The development of awareness of iron-withholding defense. *Perspectives in Biology and Medicine*. 36: 215-221.

Williamson, J.F., McLure, C.A., Baird P.N., Male, D., Millman, J., Lawley, B., Ashdown, M.L., Keating, P.J. and Dawkins, R.L. (2008) Novel sequence elements define ancestral haplotypes of the region encompassing complement factor H. Human Immunology 69: 207-219.

Williamson, J.F., Steele, E.J., Lester, S., Kalai,O., Millman, J.A., Wolrige, L., Bayard, D., McLure, C. and Dawkins, R.L. (2011) Genomic evolution in domestic cattle: ancestral haplotypes and healthy beef. Genomics 97: 304-312.

Wray, N.R., Yang,J., Hayes, B.J.,Price, A.L., Goddard, M.E. and Visscher, P.M. (2013) Pitfalls of predicting complex traits from SNPs. Nat. Rev. Genet. 14: 507-515.

Xu, M. and Cook, P.R. (2008) The role of specialized transcription factories in chromosome pairing Biochimica et Biophysica Acta 1783 : 2155-2160

Yunis, E.J., Larsen, C.E., Ferndez-Vina, M., Awdeh, Z.L., Romero, T., Hansen, J.A. and Alper, C.A. (2003) Inheritable variable sizes of DNA stretches in the human MHC : conserved extended haplotypes and their fragments or blocks. Tissue Antigens 62: 1-20.

****

Postscript

Reflections on ancestral haplotypes and citation ethics

For many years the author has had a deep fascination with the whole topic of ancestral haplotypes: and, following Ludwik Fleck, the genesis and development of ancestral haplotypes as a scientific fact has been the primary focus of this book.

When did I first know about IRF/CYO research? It was at the annual meeting of the Australian Society for Immunology in Adelaide in early December 1996 when I first became aware of "ancestral haplotypes" and their associated combinatorial "polymorphic frozen blocks". I was chairing a symposium on Multigene Families and became confused, frustrated and puzzled as I watched and listened to a superbly confident Professor Roger L Dawkins present the Perth group's latest findings on the genomic structure of the major histocompatibility complex. Surely I thought, the basic unit of inheritance was a conventional gene subject to the normal evolutionary mechanisms of natural selection and Mendelian segregation not the inheritance of a 4 Mb region of DNA encompassing almost 200 genes? While many of these genes deal with immune recognition and control of the self *versus* non-self discrimination, in both innate and adaptive immunity, there are also many genes in this region that do not have an obvious immunological role. Why are these variant genes selected and preserved within the MHC?

Over the next 15 or so years as I attended first intermittently, then more regularly, the quarterly Retreats of the Immunogenetics Research Foundation - later incorporated into the C.Y. O'Connor ERADE Village Foundation - the full import of these discoveries for the genetics and evolution of disease associations, organ and bone marrow transplantation and the quantal structure of the genome became apparent.

And then I read an interesting article in *Perspectives in Biology and Medicine* by Henry Gans with a title which explains some of my additional motivations in writing this book: "Reflections on the History and Ethics of the Proper Attribution and Misappropriation of Merit". Gans outlines how many important scientific discoveries are either deliberately overlooked or inappropriately cited by contemporaries - or even completely misappropriated by undeserving other scientists coming soon afterwards. These feelings impelled me to act given my familiarity with this important area of molecular immunology.

My primary aim has been to facilitate understanding and critically review the history and key research achievements of the IRF/CYO scientists in Perth led by Roger Dawkins since about 1983. As can be seen in this book their work in clinical and molecular immunology has been on such a scale and of such molecular precision (almost over-kill in both the quality and quantity of the data published), and appearing in many different yet high quality refereed journals that it is a real puzzle to this author, a molecular immunologist from another area of that wide discipline, why their body of work has not been more widely and *accurately* cited. Given its importance to human health it raises the question why the simple demonstration of the HGE-mediated PCR-based genomic matching technique (GMT) has not been embraced as a cheap, reliable and life saving medical diagnostic procedure particularly in bone marrow transplantation. This only added to my view that the Perth and Boston groups suffered a major scientific injustice in the post HUGO period (2000-2005). This failure of the genomics community to appropriately cite the prior discovery and characterisation of ancestral or conserved extended haplotypes over the previous 25 years did not go unnoticed.

So the scientific misappropriation that Henry Gans so well describes was well underway and overt at the turn of century. However there is the other important issue of "peer group silence" in the face of fundamental scientific discovery. It is a problem that pervades modern Science, but is of course not restricted to Science, it is both a social and psychological phenomenon associated with all professional and artistic groups - all those human endeavours that require creativity, ingenuity, focus and sheer persistence over long periods of time. It is part of the human condition. It is a characteristic our species should value highly. The discovery of ancestral haplotypes has profound implications for genetics and evolution. It is a big scientific discovery the full ramifications of which may take centuries to pervade mankind's understanding of how we have evolved and the individual structures of our genomes.

It is interesting to note that recently there has been some recognition of the importance of the Perth concepts. A Google Scholar search reveals that more appropriate citation of the earlier IRF/CYO papers is taking place in some articles published in *Nature* and *Science*. Further, in the past few years the shift to electronic publishing and electronic journal archives is making it easier for scientists and scholars to access the literature from their computers.

Further reading Postscript

Fleck, L. (1981) Genesis and development of a scientific fact. Originally published in German 1935, Translated and re-published by University of Chicago Press, Chicago and London.

Gans, H. (2011) Reflections on the history and ethics of the proper attribution and misappropriation of merit. *Persp. Biol. Med.* 54:470-478.

Gillman, M.A. (1996) Envy as a retarding force in science. Avebury Series in Philosophy. Ashgate Publishing Ltd. ISBN 1-85972 -397 -7

Schoeck, H. (1966) Envy : A theory of social behaviour. Liberty Fund. Indianapolis. ISBN 0-86597-064-5

****

Addenda

• HIV in Western Australia

Western Australia as elsewhere was not shielded from the AIDS pandemic which began in 1981. As the only medical facility set up to handle this urgent public health problem The Department of Clinical Immunology at Royal Perth Hospital became the diagnostic and coordinating centre in WA in the fight against HIV. The department coordinated with other key Australian centres in the national campaign to control the spread of the disease. Roger Dawkins convened meetings in 1982 with the heads of other clinical immunology centres in Australia, David Pennington (Melbourne), John Bradley (Adelaide), Ron Penny and Tony Basten (Sydney). They decided the best way to deal with the epidemic and submitted a plan to the Federal Minister of Health. It included protecting the blood supply, protecting hospital staff and the public and "treat rather than threaten incarceration " policy for patents suffering from AIDS. In those days the causative agent was unknown. An intensive global effort was underway to identify the aetiological agent.

The full story of this effort has yet to be told. An extended interview with Professor Dawkins by the Oral History Project of the National Library of Australia has been archived.

The Immunogenetics Research Foundation (IRF) was set up in 1983 and the research on HLA and Disease Associations somehow continued as the department grappled with the fight against HIV/AIDS. A little known fact was the contribution by Western Australian scientists to the early identification of the HIV virus the causative agent of AIDS. The work of electron microscopist Dr John Armstrong was significant. Quoted in 2008 in an interview with the medical writer Mark Thorney, Roger Dawkins said : "John Armstrong was also terrific around

Royal Perth when we were dealing with HIV and he was really the first person to see HIV in a tissue sample worldwide. He had a very disciplined approach to electron microscopy and was a huge influence on all of my immunology colleagues."

John Armstrong came to Royal Perth Hospital from the National Institute of Medical Research, Mill Hill outside London, and instituted routine lymph node biopsies as a matter of course for electron microscopic analysis. In AIDS patients the paradox of declining immunological function was associated with hyper immunological activity, lymphadenopathy - hyperplastic nodes full of mitotic activity and enlarged Germinal Centres. C-type retroviral particles were prominent features and Armstrong's work ".. demonstrated by electron microscopy of retrovirus-like particles in association with expanded follicular dendritic (FD) cells in hyperplastic lymph node follicles of homosexual males with various categories of AIDS. " And following the Perth Meetings on HLA and Disease Associations organized by Roger Dawkins in 1982 the Perth group then went on to show associations between HIV progression with HLA B35 Class I alleles particularly the 35.1/35.2 ancestral haplotypes.

It is tempting to speculate what role the fight against HIV/AIDS may have played in the research programme of IRF/CYO, particularly the latter effort throughout the 1990s when there was an intensive investigation of retroviral and retroelement sequences integrated within and around the genes within the MHC cluster. At the very least the real-world exposure to a pathogenic retroviral epidemic would have informed the formulation of what I have called the "retroviral induction model' of segmental duplication and ancestral haplotype formation .

Further reading HIV in Western Australia

Professor Roger L Dawkins interviewed by Stuart Reid, for the Oral History Project , National Library of Australia, 12th December 1993, Melaleuka Stud Canning Vale WA.

Armstrong, J.A. and Horne, R. (1984) Follicular Dendritic Cells and Viryus-Like Particles in AIDS-related Lymphadenopathy. Lancet , Aug 18, 370-372.

Armstrong, J.A., Dawkins, R.L. and Horne, R. (1985) Retroviral infection of accessory cells and the immunological paradox in AIDS Immunology Today 6:121-122.

****

• A Discussion-Interview with Roger Dawkins of the Early Years of Haplotype Research

(Synopsis of a discussion-interview between the author and RL Dawkins, May 2014)

EJS : "Can you recount for me thc key data and events leading up to, and following, that scientific meeting you held on Rottnest Island in 1982 ? "

RLD : "We were all aware from earlier studies of the MHC that raised the prospect of related genes in close linkage. It was Ceppellini who introduced the term "haplotype" to describe the alleles which segregated together in family studies (Ceppellini et al 1967). Then, over the next decade it became obvious to the HLA community that MHC haplotypes could involve multiple loci and disparate genes such as complement (Raum et al 1976; O'Neil et al 1978) and Cyp21 (congenital adrenal hyperplasia, Pollack et al 1981). It soon became clear from the work of the Boston group that these complex haplotypes were associated with diverse diseases such as insulin dependent diabetes (IDDM, Raum et al 1979) and specific and generalized autoimmune diseases (see our work in Rigby et al 1978)."

" However, by contrast, Ankylosing Spondylitis was associated with HLA-B27 rather than a haplotype (see Cross et al 1975). By 1982, opinions were divided and that was when I decided to bring together as many in the field as I could to sort out the conflicting claims (Dawkins et al 1982). Thus, were the haplotypes due to linkage disequilibrium and thus to some form of *cis* interaction presumably involving immunological pathways? Alternatively, were the haplotypes actually ancestral, continuous DNA sequences encoding disparate genes including unknowns responsible for the observed associations? "

"In order to focus on these two alternatives, we decided to use the non-committal all-inclusive term "supratype" (or "over-combination") and began the process of allocating available observations into each category or mixtures of the two (Dawkins et al 1982, 1983). The over use of the term "haplotype" was also a reason for using the non committal term "supratype" - "haplotype" as defined by Ceppellini should be reserved strictly for family studies where the linked markers and their diseases associations segregated across a three generation pedigree. We realised that linkage disequilibrium and *cis* interaction could be excluded as too simplistic descriptive terms. In our view "supratypes" were seen to be associated with the quantitative regulation of diseased states with high penetrance."

" The 1982 Rottnest meeting was crucial. We decided to go molecular and determine if supratypes could indeed be associated with a contiguous stretch of DNA. Then within a few years conclusive evidence for ancestral haplotypes was available since selected disease-associated haplotypes shared the same genomic markers implying sequence identity as distinct from *cis* interaction. We isolated by pulsed field gel electrophoresis megabase lengths of DNA sequence encoding disease associated supratype markers (Dawkins et al 1987a,b; Kay et al 1987; Saueracker et al 1987;Tokunaga et al 1988; Kay et al 1988; Tokunaga et al 1989). We were then able to confidently designate "ancestral haplotypes" at the molecular DNA level – and this was the surprizing challenge of the data (such as either 8.1, 7.1, 18.2, etc; Zhang et al 1990;Degli-Esposti et al 1992). On the other hand the Boston group called them "conserved extended haplotypes" (CEHs) from the early 1980s (Awdeh et al 1983; Raum et al 1984; Alper et al 1986;Yunis et al 2003; Alper et al 2006). "

" Importantly, we had also supplied unequivocal evidence of haplospecific regulatory genes affecting quantitative diseased-states whereby polymorphic loci within ancestral haplotypes (or supratypes) affected expression of, for example, tumour necrosis factor genes (TNF, Dawkins et al 1990a; Abraham et al 1993), insulin dependent diabetes (IDDM, Raum et al 1984; Kelly et al 1985; Dawkins et al 1990b), the serum concentration of IgA (IgAdef, Wilton et al 1985) and the rate of progression after HIV infection (Cameron et al 1990)."

EJS : " In these years were the evolutionary implications of the data becoming apparent? "

RLD : " Most certainly. We have consistently maintained that ancestral haplotypes are more than just "fixed DNA" or "conserved extended haplotypes" – in our view they are ancient haplotypes of great significance to understanding the retroviral driven evolution of the MHC which we had shown by the end of the 1990s (Dawkins et al 1999; and below). In recent years we have clearly shown that inheritance of large 'quanta" of DNA from remote ancestors is the rule rather than the exception for the wider human and mammalian genome (McLure et al 2005a,b; Williamson et al 2011; McLure et al 2013). Thus these conserved haplotypes have been inherited from remote ancestors. We used the term "ancestral" as it implies stable transmission from human ancestors at the beginning of the expansion out of Africa 50,000-100,000 years ago. The term "ancestral" is more than just speculation and is based on the wide population distribution and frequency of conserved MHC haplotypes in Caucascoids (Degli-Esposti et al 1992) and other racial/ethnic groups (Gaudieri et al 1997; Kay et al 1988; see Figure 15, Box 4 in this book). These age estimates accord well with independent calculations based on different assumptions (Smith et al 2006) and the many contributions of the Klein group on the age of MHC loci (e.g. Klein 1996)."

" It was clear to us by 1999 that the complex structural and functional features of ancestral haplotypes or conserved extended haplotypes were comprehensible in terms of imperfect segmental duplication, polymorphic frozen blocks, retroviral and retroelement indels and evolution. I should emphasise that these features have been confirmed repeatedly and they now apply generally throughout the genome."

EJS : "Yes. I would add that a key feature of the retroviral induction model is the episodic genesis of ancestral haplotypes during speciation-like events (Dawkins et a 1999). And a speciation-like event does not necessarily mean "reproductive isolation". In my reading of the literature retroelements are most often induced into their mobile mutagenic phase, by pathogenic retroviral infections which create highly mutagenic environments within cells (Chui and Greene 2005; Alysson et al 2007; Doria et al 2009; Jones et al 2013) akin to that observed in cancer cells (Lindley and Steele 2013) - involving not only error-prone reverse transcription but also cytosine-to-uracil deamination and adenosine-to-inosine RNA editing."

" I am impressed by the HLA observations in the 1990s summarised by Parham and Ohta in an excellent review. To my mind the findings point to a recent speciation–like event in the human lineage in terms of an explosion of novel inter-allelic recombinant HLA-B alleles (and by implication new ancestral haplotypes). This may have occurred during the migration and regional radiation of South American Indian tribes some 12,000 years ago (Parham and Ohta 1996). But I agree with you that in all likelihood this involved a combination of progressive replacement with concurrent retroviral infection."

References for Early Years Ancestral Haplotype Research

Abraham, L.J., French, M.A.H. and Dawkins, R.L. (1993) Polymorphic MHC ancestral haplotypes affect the activity of tumour necrosis factor-alpha Clin. Exp. Immunol. 92: 14-18

Alper, C.A., Larsen, C.E., Dubey, D.P., Awdeh, Z.L. et al. (2006). The haplotype structure of the human major histocompatibility complex. Human Immunology 67: 73-84.

Alper, C.A., Awdeh, Z.L. and Yunis, E.J. (1986). Complotypes extended haplotypes, male segregation distortion and disease markers. Hum Immunol 15: 366-373 1986.

Alysson R., Muotri, A.R., Marchetto, M.C.N., Coufal, N.G. and Gage, F.H. (2007) The necessary junk: new functions for transposable elements. Hum. Mol. Genet. 16, Review Issue 2 R159–R167.

Awdeh, Z.L., Raum, D., Yunis, E.J. and Alper, C.A. (1983) Extended HLA/complement allele haplotypes: evidence for T/t-like complex in man. Proc. Natl. Acad. Sci.(USA). 80: 259-263.

Cameron, P.U., Mallal, S., French, M.A.H. and Dawkins, R.L. (1990) MHC genes influence the outcome of HIV infection: ancestral haplotypes with C4 null alleles explain diverse associations. Hum. Immunol. 29 : 282-295.

Ceppellini, R., Curtoni, E.S., Mattiuz, P.L. et al. (1967) Genetics of leukocyte antigens: a family study of segregation and linkage. In Curtoni, E.S., P.L. Mattiuz and R.M. Tosi. (eds): Histocompatibility Testing 1967. Copenhagen, Munksgaard. pp. 149-187.

Chiu., Y-L. and Greene, W.C. (2005) APOBEC3 Cytidine deaminases: Distinct antiviral actions along the retroviral life cycle J. Biol. Chem. 281 : 8309-8312.

Cross, R.A., Rigby, R. and Dawkins, R.L. (1975) The significance of HL-A W27 in Ankylosing Spondylitis and Reiter's Syndrome with three family studies. Aust. N.Z. J. Med. 5 : 108-111.

Dawkins, R.L., Christiansen, F.T., Zilko, P.J., Carrano, J.A. et al (Eds) (1982) Immunogenetics in Rheumatology: Musculosketal Disease and D-Penicillamine. Incl D-Pen-HLA 82 Workshop, Perth, WA. Exerpta Medica, Amsterdam-Oxford-Princeton.

Dawkins, R.L., Christiansen, F.T., Kay, P.H., Garlepp, M. et al.(1983) Disease associations with complotypes, supratypes and haplotypes. Immunol. Rev. 70: 5-22.

Dawkins, R.L., Leelayuwat, C., Gaudieri, S.,Tay, G. et al. (1999) Genomics of the major histocompatibility complex: haplotypes, duplication, retroviruses and disease Immunol. Rev. 167: 275-304.

Dawkins, R.L.,Leaver, A.,Cameron, P.U., Martin, E. et al. (1990a) Some disease-associated ancestral haplotypes carry a polymorphism of TNF. Hum. Immunol. 26 : 91-97.

Dawkins, R.L., Martin, E., Saueracker, G.,Kay, P.H., et al (1990b) Supratypes and ancestral haplotypes in IDDM: Potential importance of central non-HLA MHC genes. J. Autoimmunity 3 (Supp) : 63-68.

Dawkins, R.L., Kay, P.H.,Martin, E. and Christiansen, F.T. (1987a) The genomic structure of ancestral haplotypes revealed by pulsed field gel electrophoresis. In: Immunobiology of HLA Vol I Immunogenetics and Histocompatibility. Proceedings 10th International Histocompatibility Workshop and Conference. Dupont, B. (ed). Princeton, NJ and New York City, Nov 12-23 , 1987. Springer-Verlag. 1989 pp 454-455.

Dawkins, R.L.,Tokunaga, K., Saueracker, G., Kay, P.H. et al (1987b) The MHC gene arrangements on diseases associated supratypes revealed by pulsed field gel electrophoresis. In: Immunobiology of HLA Vol II Immunogenetics and Histocompatibility. Proceedings 10th International Histocompatibility Workshop and Conference. Dupont, B.(ed). Princeton, NJ and New York City, Nov 12-23 , 1987. Springer-Verlag. 1989 pp 454-455.

Degli-Esposti, M.A., Leaver, A.L., Christiansen, F.T., Witt, C.S. et al. (1992) Ancestral haplotypes: conserved population MHC haplotypes. Hum. Immunol.34: 242-252.

Doria, M.,Neri, F., Gallo, A., Farace, M.G. and Michienzi, A. (2009) Editing of HIV-1 RNA by the double-stranded RNA deaminase ADAR1 stimulates viral infection. Nucl. Acids Res. 37 : 5848-5858.

Gaudiieri, S.,Leelayuwat, C.,Tay, G.K.,Townend, D.C. and Dawkins, R.L. (1997) The major histocompatibility complex (MHC) contains conserved polymorphic genomic sequences that are shuffled by recombination to form ethnic-specific haplotypes. J Mol Evol 45 : 17-23.

Jones, R.B., Song, H.,Xu, Y., Garrison,K.E. et al. (2013) LINE-1 retrotransposable element DNA accumulates in HIV-1-infected cells. J Virol. 87: 13307-13320.

Kay, P.H., Dawkins, R.L., Williamson, J.F., Tokunaga, K. et al. (1988) Coexistence of an MHC chromosomal segment marked by HLA B17, Bfs, C4A6, B1, DR7, and DQw9 in different ethnic groups. Hum. Immunol. 23 : 27 -35.

Kay, P.H., Williamson, J.F., Tokunaga, K., Dawkins, R.L. et al (1987) The HLA B17, BfS, C4A6, B1,DR7 and DQw3.2 supratype marks an extensive chromosomal segment shared between different ethnic groups. In: Immunobiology of HLA Vol II Immunogenetics and Histocompatibility. Proceedings 10th International Histocompatibility Workshop and Conference. Dupont, B.(ed). Princeton, NJ and New York City, Nov 12-23 , 1987. Springer-Verlag. 1989 pp 288-289.

Kelly, H., McCann, V., Kay, P., Wilton, A. et al (1985) Susceptibility to IDDM is marked by MHC supratypes rather than individual alleles. Immunogenetics 22 : 643 – 651.

Klein, J. (1996) The major histocompatibility complex and the neutral theory: the exception that proves the rule? J. Genet. 75: 117 – 124.

Lindley, R.A. and E.J. Steele, E.J. (2013) Critical analysis of strand-biased somatic mutation signatures in TP53 versus Ig genes, in genome -wide data and the etiology of cancer ISRN Genomics. Vol 2013 Article ID 921418, 18 pages.

McLure, C.A., Hinchliffe, P., Lester, S., Williamson, J.F. et al. (2013) Genomic evolution and polymorphism: Segmental duplications and haplotypes at 108 regions on 21 chromosomes. Genomics 102: 15-26.

McLure, C.A., Williamson, J.F., Smyth, L.A., Agrawal, S. et al. (2005a) Extensive genomic and functional polymorphism of the complement control proteins. Immunogenetics 57: 805–815.

McLure, C.A., Kesners, P.W., Lester, S., Male, D. et al. (2005b) Haplotyping of the canine MHC without the need for DLA typing, Int. J. Immunogenetics 32 : 407–411.

Muotri A.R., Marchetto, M.C.N., Coufal N.G. and Gage F.H. 2007 The necessary junk: new functions for transposable elements. Hum. Mol. Genet. 16: R159-R167

O'Neil, G.J., Yang, S.Y. and Dupont, B. (1978) Two HLA-linked loci controlling the fourth component of human complement. Proc. Natl. Acad. Sci. (USA)75 : 5165-5169

Parham, P. and Ohta, T. (1996) Population biology of antigen presentation by MHC class I molecules. Science. 272 : 67 -74.

Pollack, M.S., Levine, L.S., O'Neil, G.J., Pag, S. et al. (1981) HLA Linkage and B14, DRI, BfS haplotype association with the genes for late onset and cryptic 21-Hydroxylase deficiency. Am. J. Hum. Genet. 33: 540-550.

Raum, D., Glass, D., Carpenter, C.B., Alper, C.A., and Schur, P.H. (1976) The chromosomal order of genes controlling the major histocompatibility complex, properdin factor B, and deficiency of the second component of complement. J. Clin. Invest.58: 1240-1248.

Raum, D., Alper, C.A., Stein, L. and Gubbay, K. (1979) Genetic markers for insulin dependent diabetes. Lancet i 1208-1210.

Raum, D., Awdeh, Z., Yunis, E.J., Alper, C.A. and Gabbay, K.H. (1984) Extended major histocompatibility complex haplotypes in Type I Diabetes Mellitus. J. Clin. Invest. 74 : 449-454.

Rigby, R.J., Dawkins, R.L., Wetherall , J.D. and Hawkins, B.R. (1978) HLA in systemic lupus erythematosus: influence on severity. Tissue Antigens 12: 25 -31.

Saueracker, G.C.,Martin, E. and Dawkins, R.L. (1987) Some supratypes extend to DR. In: Immunobiology of HLA Vol II Immunogenetics and Histocompatibility. Proceedings 10th International Histocompatibility Workshop and Conference. Dupont, B. (ed). Princeton, NJ and New York City, Nov 12-23 , 1987. Springer-Verlag. 1989 pp 302.

Smith, W.P., Vu, Q., Li, S.S., Hansen, J.A., Zhao, L.P. and Geraghty, D.E. (2006) Toward understanding MHC disease associations: partial resequencing of 46 distinct HLA haplotypes. Genomics 87: 561-571.

Tokunaga, K., Sauracker, G., Kay, P.H., Christiansen, F.T. et al. (1988) Extensive deletions and insertions in different MHC supratypes detected by pulse field gel electrophoresis. J. Exp. Med. 168: 933-940.

Tokunaga, K., Kay, P.H., Christiansen, F.T., Saueracker, G. and Dawkins, R.L. (1989) Comparative mapping of the human major histocompatibility complex in different racial groups by pulsed field gel electrophoresis. Hum. Immunol. 26 : 99-106.

Wilton, A., Cobain, T.J. and Dawkins, R.L. (1985) Family studies of IgA deficiency. Immunogenetics. 21 : 333-342.

Williamson, J.F., Steele, E.J., Lester, S., Kalai, O. et al. (2011) Genomic evolution in domestic cattle: ancestral haplotypes and healthy beef. Genomics 97: 304-312.

Yunis, E.J., Larsen, C.E., Ferndez-Vina, M., Awdeh, Z.L. et al. (2003) Inheritable variable sizes of DNA stretches in the human MHC : conserved extended haplotypes and their fragments or blocks. Tissue Antigens 62: 1-20.

Zhang, W.J., Degli-Esposti, M.A., Cobain, T.J., Cameron, P.U. et al. (1990) Differences in gene copy number carried by different MHC ancestral haplotypes: quantitation after physical separation of haplotypes by pulsed field gel electrophoresis. J.Exp.Med. 171 : 2101-2114.

List of co-authors on IRF/CYO papers

Abraham LJ
Agrawal S
Akinshipe BO
Alarcon-Segovia D
Albert ED
Alper C
Amadou C
Anand R
Andreas A
Andreas-Zietz A
Arimori S
Armstrong JA
Arnaiz-Villena A
Ashdown ML
Awad J
Bach JF
Baird PN
Baker D
Balmern L
Bardy P
Bayard D
Bellgard MI
Bernoco D
Bernoco D
Berry J
Bonifacio E
Bontrop R
Boongird P
Bowling AT
Bradley B
Bradley J
Brown J
Bucknall R
Burger HG
Caillat-Zucman S
Calin A
Cambon-D/U
Thomsen A
Cameron P
Cameron PU
Campbell RD
Carcassi C
Carnegie P
Carrano JA
Chan SH
Chan-Lui WY
Chapman J
Charoenwong P
Charron D
Cheah PS
Chen R
Chiewsilp P
Chin DU
Chistiansen FT
Ciccone E
Ciccone E
Clark WT
Clayton G
Cobain TJ
Cohen ML
Contu L
Curtoni S
Dallas P
Daly LN
Davies RA
Dawkins B
Dawkins JR
Dawkins ST
Degli-Esposti MA
Denis KA
Dipeveen D
Dunham I
Edward E
Eghtedari A
Erlich HA
Farrow B
Fauchet R
Feeney D
Feeney J
Feltkamp TEW
Festenstein H
Finlay JM
Forrester DM
French MAH
Froudist J
Garlepp MJH
Gaudieri S
Gedeon A
Genkins G
Giffiths M
Giles KM
Gillieat SJ
Gin W
Giphart M
Goatcher P
Gofton P
Gojobori T
Griffiths MS
Grimsley G
Gyllensten UB
Habara K
Hanrahan P
Hauptmann G
Hawkins BR
Hayakawa J
Heise ER
Herrmann RP
Hiew HL
Hockey A
Holborow EJ
Hollingsworth PN
Holmes K
Homma M

Horne R
Horton B
Houliston JB
Hui J
Illeni MT
Imanishi T
Inoko H
Ip M
Jackson AN
James IR
Jeannet M
Jones PP
Jonker M
Joseph GA
Juji T
Kalai O
Kalden JR
Kalden JR
Kay PH
Keating PJ
Keenan B
Keller E
Kelly H
Kemp A
Kenter M
Kesners PW
Ketheesan N
Kinnear BF
Kinnear BF
Kirk RL
Kitchen VS
Klein J
Komori K
Kulski JK
Langlands DR
Laird R
Lam K
Lambert PH
Lange A
Lattimore M
Lawley B
Lawlor D
Lawton J
Leaver AL
Leedman RJ
Leelayuwat C
Lernmark A
Lester S
Lindley RA
Lindstrom J
Lloyd SS
Loiseau P
Longman-Jacobsen N
Luciani G
Ma J
Mackay IR
Maini RN
Major GAC
Male D
Mallal S
Mallon DFJ
Mann AJ
Manning LS
Marley J
Marley JV
Marshall B
Martin E
Martinez OP
Martinez P
Maruya E
Marzo A
Mastaglia FL
Masters CL
Matz LR
Mauff G
McCann VJ
McCleod J
McLure CA
McCluskey J
McDonald BL
McDougal JS
McDuffie FC
Mehra NK
Millman JA
Moretta L
Moriuchi J
Morley AA
Muller C
Naito S
Natsuume-Sakai S
Nedospasov SA
Nerl C
Nishimukai H
Norberg R
O'Connor KJ
O'Neill GL
Owen ET
Padadimitriou JM
Pain GN
Panegyres PK
Parham P
Partanen J
Patel Y
Peers J
Pellegris G
Pende D
Pende D
Penhale WJ
Penn A
Penny R
Perstell R
Peter JB
Peter JB
Petersdorf EW
Pinelli M
Pollack MS
Potter RM
Pulendran B
Ray PH
Richardi P
Richmond J
Rigby RJ

Rischmueller M
Robinson J
Robinson W
Rodger B
Saji H
Salmon B
Sargent CA
Saueracker G
Schalke B
Schmitt LH
Schon-Hegrad M
Shaw SE
Shearwood W
Simons MJ
Simpson JA
Smith LK
Smyth LA
Spies T
Steele EJ
Stewart BJ
Stewart C
Stuckey MS
Suzuki K
Tabarias HA
Tait BD
Tan CB
Tan EM
Tanaka H
Tay KG
Taylor E
Teisserenc H
Thevasthasan M
Thomas R
Tiilikainen A
Tindall RSA
Tokunaga K
Townend D
Tsuji K
Uko G
van Rood JJ
Viale O
Viale O
Vincent A
VisudhiphanP
Wakeland EK
Warlow R
Warren HS
Watanabe H
Watkins DI
Welborn TA
Wetherall JD
White P
Whitehead AS
Whittingham S
Williamson JF
Wilson G
Wilson M
Wilton A
Witt CS
Wolrige L
Wong V
Wood PHN
Woolcock AJ
Wu X
Yang J-S
Yeung R
Yu YL
Zaledi K
Zhang WJ
Zilko PJ

• Some Visitors to IRF/CYO since 1990

1 = Major contribution to laboratory development
2 = Major contribution to concepts

1990-1991
[1]Prof. AS Whitehead (Division of Immunology, Children's Hospital and Department of Pediatrics, Harvard Medical School, Boston, USA)
Prof. A Morley (Department of Haematology, Flinders Medical Centre, SA)
[1,2] Dr M Simons (GeneType Pty Ltd)
Dr P Cameron (Department of Clinical Immunology RPH and AIDS Pathogenic Research Unit, Macfarlane Burnet Centre for Medical Research, Fairfield, Vic)
[2] Dr RM Steinman (Laboratory of Cellular Physiology and Immunology, Rockefeller University, New York)
Prof. L Weitkamp (Rochester University, New York)
[1,2] Prof. M Giphart (Leiden University Hospital, The Netherlands)

1992-1993
Dr S Vaidya (University of Texas)
Dr J Capper (University of California)
[1,2] Dr M Simons (GeneType Pty Ltd)
[1] Mr J Corbett Jnr (Corbett Research, Sydney)
[1,2] Prof. M Giphart (Leiden University Hospital, The Netherlands)
[1,2] Dr M Gassler (Leiden University Hospital, The Netherlands)
Dr P Chiewslip (Tamathibodi Hospital, Thailand)
Dr M Varney (Royal Melbourne Hospital)

Dr L Weitkamp (Rochester University, New York)
Dr B Maher (Department of Haematology, Royal Perth Hospital)
Dr J Wetherall (Curtin University, WA)
Dr R Holdsworth (Blood Transfusion Service Melbourne)
Dr S Sevall (Specialty Labs , Los Angeles)
Dr B Linn (Vision Systems, South Australia)
Dr A Morley (Flinders Medical Centre, South Australia)
Dr G Sutherland (Adelaide Children's Hospital)
Dr T Orpin (Applied Biosystems Australia)
Prof. I McKenzie (Austin Research Institute, Melbourne)
Dr G Yeoh (University of Western Australia)
Dr HS Warren (Woden Valley Hospital, Canberra)
Dr B McDonald (Prince of Wales Hospital, NSW)
Dr T Kyne (Walter and Eliza Hall Institute, Melbourne)
Dr K Johnson (Orbital Engine Corporation)
Prof. P Carnegie (Murdoch University)
Dr M Jones (Murdoch University)
Dr R MacDonald (AMRAD Pty Ltd)
Dr B Tait (Royal Melbourne Hospital)
Dr T Ratajczac (Queen Elizabeth II Medical Centre, Perth)
Dr C Davies (Wageningen University, The Netherlands)

1993-1994
Dr T Ratajczak (Sir Charles Gairdner Hospital, Perth)
[1,2] Dr M Simons (GeneType Pty Ltd, Melbourne)
[1,2] Dr M Gassler (AZL, Leiden, The Netherlands)
Dr R MacDonald (AMRAD, Pty Ltd Melbourne)
Mr R Klupacs (AMRAD, Pty Ltd Melbourne)
Dr D Hurley (Royal Perth Hospital, Perth)
Dr M Hattori (Joslin Diabetes Centre, Boston)
Dr J Cannon (Dept. of Chemistry, Murdoch University)
Dr J Armstrong CALM, Perth)
Dr C Field (Institute for Genomic Research)
Dr T Imanishi (National Institute of Genetics, Japan)
Dr K Bhatia (Goroka, PNG)

Dr M Beilhertze (Dept Microbiology, Sir Charles Gairdner Hospital, Perth)
Dr X Gao (John Curtin School of Medical Research, ANU)
Dr S Breit (St Vincent's Hospital, Sydney)
Dr B Morris (University of Sydney)
Dr A Lopez (The Hanson Centre for Cancer Research, Adelaide)
Dr J Hetzel (Division of Tropical Animal Production, CSIRO, Queensland)
Dr T Radford (AMRAD Pty Ltd, Melbourne)
[1,2] Dr P Cameron (The McFarlane Burnett Institute, Melbourne)
Dr J Mathews (Menzies School of Health Research, Darwin)
[1,2] Dr M Radman, (Institute Jacques Monod, Universite de Paris, France)
Dr S Humphries (University College London Medical School, London, UK)
Mr J Corbett Snr (Corbett Research, Sydney)
Mr T Coulepis (AMRAD Pty Ltd, Melbourne)
Mr J Hodder (AMRAD Pty Ltd , Melbourne)
[1,2] Prof. M Giphart (AZL, Leiden, The Netherlands)
Dr L Harrison (Walter and Eliza Hall Institute, Melbourne)
Dr J Wetherall (Curtin University, WA)
Dr A Wilks (Ludwig Institute for Cancer research, Melbourne)
Dr A Ziegler (Universitatsklinikum Rudolf Virchow Spandauer Damm, Berlin, Germany)

1994-1995
Prof. K Muller (President Royal College of Pathologists Australasia)
Prof. M Giphart (AZL, Leiden, The Netherlands)
Prof. T Speed (Monash University, Melbourne)
[1] Mr J Corbett Snr (Corbett Research, Sydney)
Dr J Armstrong (CALM, Perth)

Mr J Hodder (AMRAD Pharmacia Biotech, Melbourne)
Dr S Carrol (AMRAD Pharmacia Biotech, Melbourne)
Dr P Cameron (The MacFarlane, Burnet Centre for Medical Research, Melbourne)
Mr D Hilditch (Graduate School of Management, UWA)
Prof J Mattick (Centre for Molecular Biology & Biotechnology, University of Queensland)
Dr A Wilks (Ludwig Institute for Cancer Research, Melbourne)
[1] Mr John Corbett Jnr (Corbett Research, Sydney)
Mr A Roberts (Multiplex Constructions, Perth)
M G Eaton (Multiplex Constructions, Perth)
Prof. J Papadimitriou (University Department of Pathology, Queen Elizabeth II Medical Centre, Perth)
Dr J Harvey (University Department of Pathology, Queen Elizabeth II Medical Centre, Perth)
Dr P Heenan (Department of Cutaneous Pathology, Perth)
Dr Peter Leedman (Department of Medicine, Royal Perth Hospital)
[1,2] Dr M Simons (GeneType diagnostics Pty Ltd, Melbourne)
Mr C Goodhew (Pizzey & Co , Melbourne)
Mr J Pizzey (Pizzey & Co , Melbourne)
Dr R Slade Department of Zoology, University of Queensland)
A/Prof. B Ward (Murdoch University, WA)
Prof. P Thompson ((University Department of Medicine, Queen Elizabeth II Medical Centre, Perth)
Mr N Rothnie (CALM, Perth)

1995-1996

Dr N Saksena (Department of Virology, Westmead Hospital, Sydney)
Dr P Cameron (The MacFarlane, Burnet Centre for Medical Research, Melbourne)
Mr Val McKelvey (His Majesty's Theatre, Perth)
Dr M Bellgard, Computer Sciences, Murdoch University, WA)

[1,2] Dr Y Kulski (Department of Microbiology, Royal Perth Hospital, Perth)
Dr D Hilton (AMRAD Pty Ltd, Melbourne)
[1,2] A/Prof. M Chorney (Penn State University College of Medicine, Hershey Medical Centre, USA)
A/Prof. P. Mongkolwongrol (Faculty of Associated Medical Science, Khon Kaen University, Thailand)
A/Prof. G Lertmemongkolchai (Faculty of Associated Medical Science, Khon Kaen University, Thailand)
[1,2] Prof. P. Pontarotti (Centre National de la Recherche Scientifique, Toulouse, France)
[1,2] Dr C O'hUigin (Max Plank Institute for Biologie, Tubingen, Germany)
Dr T Littlejohn (Australian National Genomic Information Service (ANGIS), Sydney)
Dr P Panegyres (Department of Neuropathology, Royal Perth Hospital, Perth)
Dr F van Bockxmeer (Department fo biochemistry, Royal Perth Hospital, Perth)

1996-1997
[2] Prof K Tsuji (Japan)
Prof. P Chiewslip (CKICC HIV Centre , Khon Kaen University, Thailand)
[1,2] Dr C Leelayuwat (CKICC HIV Centre , Khon Kaen University, Thailand)
Prof AZM Zaini (University of Malaysia Transplantation Centre, Malaysia)
Prof. J Bosco (University of Malaysia Transplantation Centre, Malaysia)
Dr B Tait (Royal (Melbourne Hospital)
Dr A Geczy (University of Sydney)

1997-1998
Dr W Kasinrerk (Department of Clinical Immunology, Faculty of Associated Medical Sciences, Chiang Mai University, Thailand
[1,2] Dr C Leelayuwat (Khon Kaen University, Thailand)
Dr G Chelvanayagam (Australian National University)

[1,2] Dr S Gaudieri (Center for Information Biology, National Institute of Genetics, Mishima, Japan)
Dr M Phipps (Malaysia) (University of Malaysia Transplantation Centre, Malaysia)
Dr C Fisher (Fisher Biotech, WA)
Mr D Hilditch (Graduate School of Management, UWA)

1998-1999
Prof. P Chiewsilp (CKICC HIV Centre , Khon Kaen University, Thailand)
[1,2] Dr C Leelayuwat (Khon Kaen University, Thailand)
[1, 2] Dr S Gaudieri (National Institute of Genetics, Mishima, Japan)
Dr M Phipps (Malaysia)
[2] Dr L Smyth (University of Notree Dame WA)
Prof J Penhale (Murdoch University, WA)
Mr C Fisher (Fisher Biotech Pty Ltd, WA)
[1,2] Dr M Whisson (Red Cross Blood Transfusion Service, Perth)

1999-2000
[2] Prof J Epplen (Ruhr-Universitat Bochum, Germany)
[2] Prof. C Yung Yu (Ohio State University, USA)
[1, 2] A/Prof C Leelayuwat (Khon Kaen University, Thailand)
Mr C Fisher (Fisher Biotec, Perth)
Prof W Baily (Department of Zoology, University of Western Australia)
Prof A Bittles 9Edith Cowan University, Perth)
Dr A Barton (Department of Chemistry, Murdoch University, WA)
M E Saba (Duak Access, Perth)
Prof T Speed (Walter and Eliza Hall Institute, Melbourne; Yale University, New Haven)
Dr J Craig (Murdoch Institute, Melbourne)
Mr John Corbett (Corbett research, Sydney)
Dr J McCarthy (Fremantle Hospital, WA)
Dr R Hopkins (Fremantle hospital, WA)

A/Prof D Sampson (Dept. Electrical and Electronic Engineering, University of Western Australia)
[1,2] Prof P Pontarotti (INSERM, Marseille, France)
[1,2] Prof T Gojobori (National Institute of Genetics, Mishima, Japan)
Dr Ikeo (National Institute of Genetics, Mishima, Japan)
Mr Fukuma (Hitachi, Japan)

2000-2001
Prof D Martin (Victor Chang Cardiac Research Institute, Sydney)
[1,2] A/Prof EJ Steele (John Curtin School of Medical Research, ANU)
[1,2] Dr L Abraham (School of Biomedical, Biomolecular and Chemical Sciences, The University of Western Australia)
[1,2] Dr P Martinez (School of Pathology and Laboratory Medicine, The University of Western Australia)
A/Prof JD Wetherall (Curtin University, WA)
Prof P Carnegie (Murdoch University, WA)

2001-2002
7th International Workshop on MHC Evolution & VIII Complement Genetics Workshop and Conference- Organized and run by IRF/CYO group at CYO Canning Vale, Noalimba Conference Centre and Fremantle Town Hall (Opening) and Conference Dinner (Fremantle Gaol) -6th to 11th October 2002.
[1,2] Amadou C (University of Texas, USA)
Anzai T (Tokai University School of Medicine, Japan)
Arnaiz-Villena A (Universisdad Complutense, Madrid, Spain)
Belov K (Australian Museum, Sydney)
Birmingham DJ (Ohio State University, USA)
Bos DH (Univesrity of Canterbury, Christchurch, NZ)
Boyd GW (University of Tasmania)
Brahram S (Louis Pasteur University, Strasbourg, France)
Browning TL (Macquaria University, Sydney)

Cadavid LH (University of New Mexico, Albuquerque, USA)
Christadoss P (University of Texas Medical Branch, Galveston, USA)
De Tomaso AW (Department of Pathology, Stanford University School of Medicine, USA)
Debenham S (Animal Health Trust, Suffolk, UK)
Dijkstra JM (National Research Institute of Aquaculture, Tamaki, Japan)
Due M (University of Southern Denmark, Odense, Denmark)
Fischer Lindahl K (University of Texas, USA)
Fugita T (Fukushima Medical University, Japan)
Gao X (NCI, Frederick, Maryland, USA)
Hashiba K (Saitama Prefectural University, Saitama, Japan)
[1] Inoko H (Tokai University School of Medicine, Japan)
[1,2] Kasahara M (Graduate University for Advanced Studies, Kanagawa, Japan)
Kumanovics A (Southwestern Medical School, University of Texas, Dallas, USA)
Mack M (Institute (Institute of Legal Medicine, Mainz, Germany)
Magor K (University of Alberta, Canada)
Miller RD (University of New Mexico, Albuquerque, USA)
Noroaka N (Graduate School of Science, University of Tokyo, Japan)
Nyholm SV (Department of Pathology, Stanford University School of Medicine, USA)
Ohta Y (Department of Microbiology & Immunology, University of Maryland, USA)
[1,2] Oppenheimer S (Green College, Oxford)
Rahman A (Umea University, Sweden)
Rand V (Wellcome Trust Sanger Institute, Hixton, UK)
Renard C (INRA Jouy en Josus, France)
Rogers S (Institute for Animal Health, Berks, UK)
Satta Y (Graduate University for Advanced Studies, Japan)

[1,2] Schneider PM (Johannes Gutenberg Universitat, Maiz , Germany)
[1] Shiina T (Tokai University School of Medicine, Japan)
[1,2] Steele EJ (John Curtin School of Medical Research, ANU)
Suzuki T (Graduate University for Advanced Studies, Hayama, Japan)
Takahashi T (Graduate University for Advanced Studies, Kanagawa, Japan)
Thomson G (Victor Chang Cardiac Research Institute, Sydney)
Turner TF (University of New Mexico, Albuquerque, USA)
Walker BA (Institute for Animal Health, Berks, UK)
Walker I (University of Gottingen, Germany)
[1,2] Yu CY (Ohio State University, USA)
[2] Ziegler A (Humboldt-Universitat, Berlin, Germany)

2003-2004
Dr R Hahnel (Royal Perth Hospital, UWA)
Dr M Ashdown (Genetic Technologies Ltd, Melbourne)
Dr L Ashdown (Genetic Technologies Ltd, Melbourne)
P Coombes P (Environmental Engineering)
Mr J Mensink (Premier's Department, WA State Government)
Mr A McTaggart (Department of Environment, WA State Government)
Dr S Shea (University Notre Dame, Perth)
[1,2] Dr L Smyth (University Notre Dame, Perth)
[1,2] Ms S Lester (Hanson Institute –IMVS, Adelaide)

2004-2005
Prof J-S Yang (National Taiwan Ocean University)
Prof C-H Wu (National Taiwan Ocean University)
Dr S Shea (University of Notre Dame, Perth)
Mr T Duncanson (Oil Mallee Company, WA)
[1,2] Ms S Lester (Hanson Institute-IMVS, Adelaide, SA)
Dr M Flocco (Bruker Daltonics, Leipzig, Germany)
[1,2] Prof S Oppenheimer (Green College, Oxford)

Mr J Pfistershammer (Ag ID, WA)
Mr G Holmes (Tanks West, WA)
Dr A Montoya (University of Notre Dame, Perth)
Mr D Singh (Prime Solar Ltd, WA)
[1,2] Dr M Whisson (Red Cross Blood Bank, Perth)
Dr A Veitch (Veterinary Surgeon, WA
[1,2] Dr C Amadou (University of Texas, USA)
Dr M Jacobson (Genetic Technologies Ltd, Melbourne)
Dr A Hodgson (Genetic Technologies Ltd, Melbourne)
[1,2] Dr MJ Chorney (Department Microbiology and Immunology, Penn State University, USA)

2005-2006

Dr JM Angles (Animal Referral Hospital, Sydney)
[1,2] Dr MJ Chorney (Department Microbiology and Immunology, Penn State University, USA)
[2] Dr LJ Kennedy (Centre Integrated Genomic Medical Research, University of Manchester, UK)
Dr G Cobon (Genetic Technologies Ltd, Melbourne)
Prof G Wilcox (Murdoch University, WA)
Mr S Halse (Department of Conservation and Land Management, Science Division,WA)
[1,2] Ms S Lester (Hanson Institute –IMVS , Adelaide)
[1,2] Dr L Smyth (University of Notre Dame, Perth)
Dr L Ashdown (Genetic Technologies Ltd, Melbourne)
Mr J Pfistershammer (Ag ID, WA)
Mr T Clarke (SVA Motor Vehicles)
Prof P Panegyres (University of Western Australia)

2006-2007

[1,2] Ms S Lester (Hanson Institute-IMVS, Adelaide)
[1] Dr JM Angles (Animal Referral Hospital, Sydney)
[1] Dr S Holloway (Department of Veterinary Clinic and Hospital, University of Melbourne)
Dr M Ashdown (Genetic Technologies Ltd, Melbourne)
[1,2] Dr M Whisson (Red Cross Blood Bank, Perth)

2007-2008

Mr T Schiavello (Qiagen Pty Ltd)

Prof J-S Yang (National Taiwan Ocean University)
[1,2] Prof M Degli-Esposti (Lions Eye Institute, Perth)
[1,2] Ms S Lester (Hanson Institute-IMVS, Adelaide)
Dr R Staaden (Veterinary Surgeon, WA)

2008 -2009

Dr I Edwards (Edstar Genetics Pty Ltd, WA)
Dr M Trudgeon (University Of Western Australia)
[1,2] Dr M Whisson (Red Cross Blood Bank, Perth)
[1,2] Dr P Hollingsworth (University of Western Australia)
[1,2] Dr L Smyth L (University of Notre Dame, Perth)
[1,2] Prof P Panegyres (University of Western Australia)
Prof J Pluske (Animal Research Institute, Murdoch University, WA)
Dr L Ashdown (Genetic Technologies Ltd, Melbourne)
Dr S Holloway (Department of Veterinary Clinic and Hospital, University of Melbourne)
Mr S Halse (Department of Conservation and Land Management, Science Division, WA)
Fenny R (Maitraya, Luxury Private Retreat & Rick Fenny Group, WA)
[1,2] Bayard D (Global Reproductive Solutions Pty Ltd, Victoria)
Pfistershammer J (Ag ID, WA)

2009-2010

[1,2] Steele EJ (Melville Analytics Pty Ltd, Wollongong)
[1,2] Lindley RA (Melville Analytics Pty Ltd,Wollongong)
[1,2] Bayard D (Global Reproductive Solutions Pty Ltd, Victoria)
Mr A Peggs (WR Carpenters Pty Ltd, WA)
Mr J Emanuel (Emanuel-Dawes Pastoral, WA)
Prof G Wilcox (Murdoch University, WA)
Dr R Pashen (Veterinary Surgeon, Vic)
A/Prof Groth D (Curtin University, WA)
A/Prof JD Wetherall (Curtin University, WA)
Dr L Ashdown (Genetic Technologies Ltd, Melbourne)
Dr I Edwards (Edstar Genetics Pty Ltd, WA)
The Facey Group (WA)

Dr SS Lloyd (Mediscan, Canning Vale)

2010-2011 (Includes Retreat on "The World" organised by RL Dawkins, P Hinchliff)
Prof M Borowitzka (Murdoch University, WA)
[1,2] Mr D Bayard (Global Reproductive Solutions Pty Ltd, Victoria)
[1,2] Prof M Radman (INSERM Necker Hospital, Paris, France and Mediterranean Institute Life Sciences MedILIS , Split, Croatia)
Prof E Friedberg (Co-Director MedILIS and Southwestern Medical Center, University of Texas, Editor *DNA Repair*)
Mr A Brown (Author& Journalist, *The Guardian, The Sunday Times*)
[1,2] Nix V (Cattle Farmer, WA)
[1,2] Ms S Lester (Hanson Institute –IMVS, Adelaide)

2012-2013
[1,2] A/Prof JK Kulski (University of Western Australia)
[1,2] Dr N Jacobsen (University of Western Australia)
[1,2] Dr EJ Steele EJ (Melville Analytics Pty Ltd, Jindabyne, NSW)
Dr P McGilchrist (Murdoch University, WA)
A/Prof W Greene (Murdoch University, WA)
Dr K Oliver (Murdoch University, WA)
Scadding C (TWS Analytics, Centre Forensic Science , UWA)
Dr M Brandon (Genetic Technologies Ltd , Melbourne)
Dr L Ashdown (Genetic Technologies Ltd , Melbourne)
Dr L Smyth (University of Notre Dame, Perth)
Mr A Peggs (Alan Peggs Rural Pty Ltd, WA)
[1,2] Mr D Bayard (Global Reproductive Solutions Pty Ltd, Victoria)
[1,2] Ms S Lester (Hanson-IMVS, Adelaide)
Mr R Candy (Erwin Hunter Co, WA)
Mr S Cook (Gallagher Animal Management Systems, WA)
Mr V Garreffa (Mondo Butchers, WA)

Mr M Cabassi (Stock Feeder, AI Specialist, WA)
Dr P Trenchev (Specialist Pathologist, Kemps Creek, NSW)
[1,2] Nix V (Cattle Farmer, WA)
Manning G (Local Dairy Farmer, Serpentine, WA)

Glossary Boxes 2, 5-9

Box 2 The **Hardy–Weinberg principle** (also known as the **Hardy–Weinberg equilibrium, model, theorem**, or **law**) states that allele and genotype frequencies in a population will remain constant from generation to generation in the absence of other evolutionary influences. These influences include *mate choice*, *mutation*, *selection*, *genetic drift*, *gene flow* and *meiotic drive*. Because one or more of these influences are typically present in real populations, the Hardy–Weinberg principle describes an ideal condition against which the effects of these influences can be analyzed.

In the simplest case of a single locus with two alleles denoted A and a with frequencies $f(A) = p$ and $f(a) = q$, respectively, the expected genotype frequencies are $f(AA) = p^2$ for the AA homozygotes, $f(aa) = q^2$ for the aa homozygotes, and $f(Aa) = 2pq$ for the heterozygotes. The genotype proportions p^2, $2pq$, and q^2 are called the Hardy-Weinberg proportions. [Note that the sum of all genotype frequencies of this case is the binomial expansion of the square of the sum of p and q, and such a sum, as it represents the total of all possibilities, must be equal to 1. Therefore $(p + q)^2 = p^2 + 2pq + q^2 = 1$. The solution of this equation is $q = 1 - p$.
If union of gametes to produce the next generation is random, it can be shown that the new frequency f' satisfies $f'(A)= f(A)$ and $f'(a)= f(a)$

That is, allele frequencies are constant between generations.

This principle was named after G. H. Hardy and Wilhelm Weinberg, who first demonstrated it mathematically.

Note that p and q can refer to the frequencies in the population of long extended haplotypes such as ancestral haplotypes (not just simple alleles of genes) showing that these are assorting randomly.

Box 5

Retro-like elements (RLEs) and Transposable elements (TE) in the Human Genome

• DNA Transposons -2.8%
Currently not active; active in early primate evolution to ~37mya

• LTR Retrotransposons - 8.3%
Long terminal repeat
Human LTR elements are endogenous retroviruses (HERVs). Most inserted >25mya activity currently limited.

• Non LTR Retrotransposons

LINE -16.9% Mainly L1, long interspersed nuclear elements, currently active, arose ~150mya.

Alu -10.6% (also classed as SINEs, short interspersed nuclear elements); currently active, arose ~65mya ; subfamilies used as molecular 'time' markers to study evolution gene systems. Generation polymorphism via microsatellite/dA expansion (middle A-rich and 3'PolyA tail).

SVA -0.2% Recombinant SINE, variable number tandem repeats, Alu-like region; arose 25mya still active

Others-6.0%

• Non-Transposable Elements in the Human Genome ~ 55%

Cordaux, R. and Batzer,M.A. 2009 The Impact of Retrotransposons on Human Genome Evolution Nat Rev Genet 10: 691-703

Box 6

LINE - 1 (L1 elements) - non-LTR retrotransposons

• >500,000 copies in the human genome

• Continued mobilisation activity for past 150 Mya

• Only about 100 still active - *in vitro* cell cultures can detected these active ones - may also be activated in cancer genomes. Activation signals?: stress, retroviral and other viral infections, and inflammatory signals

• Internal RNA Pol II promoter

• They drive mobility of Alu elements (i.e. provide reverse transcriptase activity, RT)

• Canonical full length L1 element ~ 6Kb

• Located in gene-rich regions

• Located in intervening regions in the main (lethal insertional mutants in protein-coding and regulatory regions selected against - but some clear diseases associated with insertional mutagenesis)

Cordaux,R.and Batzer, M.A.(2009) The Impact of Retrotransposons on Human Genome Evolution Nat Rev Genet 10: 691-703

Box 7

Alu elements

• ~1,100,000 copies in human genome

• Continued mobilisation activity for past 60 Mya

• Some younger Alus still active - driven by L1 reverse transcriptase activity. Activation signals?: stress, retroviral and other viral infections, and inflammatory signals

• Internal RNA Pol III promoter

• Canonical full length Alu element ~ 300bp

• Dimeric structure (7SL RNA gene-signal recognition particle)

• Located in gene-rich regions

• Many (70%) located in intronic regions (as with L1 elements lethal insertional mutants in protein-coding and regulatory regions selected against in evolution- but some clear diseases associated with insertional mutagenesis)

Cordaux, R. and Batzer, M.A.(2009) The Impact of Retrotransposons on Human Genome Evolution Nat Rev Genet 10: 691-703

Box 8

Alu elements - Evolutionary Markers

Based on DNA consensus, Alu sequences an be divided into at least three families (and their subfamilies):

• Oldest Alu J - expanded 81 million years ago

• Primate specific Alu S
- sub family S (x,q,p,c,; platyrrhine-specific, expanded 35-44 mya, and
- Alu Sg about 31 mya

• Primate specific Alu Y subfamily 19 mya to several hundred thousand years ago (catarrhine-specific)

• Human specific Alu Ya and Yb 6-7 mya - some still active

Arcot, S.S., Adamson, A.W., Lamerdin, J.E.,Kanagy, B., Deininger, P.L, Carrano, A.V. and Batzer, M.A. (1996) Alu fossil relics- distribution and insertion polymorphism Genome Res. 6 : 1084-1092

Kapitonov, V. and Jurka, J. (1996) The age of Alu subfamilies J Mol Evol 42: 59-65.

Cordaux,R.and Batzer, M.A.(2009) The Impact of Retrotransposons on Human Genome Evolution Nat Rev Genet 10: 691-703

Box 9

HERV Human Endogenous Retrovirus

• ~200,000 copies in human genome

• Most inserted >25 Mya

• Most degraded and inactive in humans

• LTRs and RNA Pol II promoters within LTRs

• Canonical full length HERV element ~ 9000bp but usually smaller and fragmented

• Theoretical spacing ~ 10 Kb on average.

• Again located in intervening regions in main (as with L1 elements lethal insertional mutants in protein-coding and regulatory regions selected against in evolution- but some clear diseases associated with insertional mutagenesis)

Cordaux R & Batzer MA 2009 The Impact of RetroTransposons on Human genome Evolution Nat Rev Genet 10: 691-703

Index

Printed in Dunstable, United Kingdom

91226145R00116